Time:
The Theory of Everything

by

Dr. Lee

DORRANCE
PUBLISHING CO
EST. 1920
PITTSBURGH, PENNSYLVANIA 15238

Dorrance Publishing Co
585 Alpha Drive
Suite 103
Pittsburgh, PA 15238
Visit our website at *www.dorrancebookstore.com*

ISBN: 978-1-6453-0597-2
eISBN: 978-1-6453-0615-3

Introduction

This is a book about science and scientific theory. However, it is not written for scientists. It is for the general public who have an interest in science — the ones who watch those television shows or read articles or books about the universe, time travel, black holes, and the like. The one thing in common with all these shows, publication, and articles is, at the end of all of them, the scientists usually say they don't have the answers to the questions they have posed in the article or show. Either they don't understand why something happens, or they don't know what something is or what's causing something.

This book will try to answer some of these questions. I will also try to explain or answer some of the paradoxes the theories of time travel, black holes, multiple universes, and the like pose. This book is not filled with mind-numbing calculations, and I try not to use too much scientific jargon, but sometimes it is unavoidable. It's written in simple English. I simply look at science logically and without implying any far-out theories or using any unproven facts. I use only the numbers today's scientists use and the rules of science as they are taught today in universities throughout the world.

As you go through this book, I will point out the paradoxes some of the theories imply and debunk other theories completely. Many of them are right out of *Star Trek* — they either cannot be proven because there is no way to test them, or they ignore certain facts or rules that don't quite fit their theory. Some are even unproven theories based on other unproven theories.

Hopefully by the end of this book, you will have a whole new perspective of science and scientists and understand the universe as you never have before. And most of all, I hope to demystify cosmic science and make it accessible to everyone.

Preview

If you ask any astronomer or physicist how old the universe is, they will tell you it's about 13.7 billion years old. The science that determines this is simple: light has a set speed, and if we know how far away we can see, we know how much time it took light to travel that distance. It is simple math. On that we can all agree.

But when you ask how big the universe is, problems begin. We can see 13.7 billion light-years in one direction and 13.7 light-years in the opposite direction. So that makes the universe at least 27.4 billion light-years across. So what's the problem?

This assumes there is nothing beyond the farthest galaxy we can see, as if to say, if you are on a planet in that galaxy 13.7 billion light-years away and look forward, you will see nothing. That galaxy is essentially leading the parade of other galaxies as they fly off into the unknown of deep space. We'll see about that later.

Science also states when you investigate space, you are looking back in time. The deeper into space you look, the farther back in time you are looking. Yet when today's scientists talk about galaxies light-years away, they say that galaxy *is* going away from us at X speed depending on how many billions of light-years away it is when they should be saying that galaxy *was* going away from us at X speed. Now this may seem like a small detail, but it changes everything. The difference leads science to come up with things like dark energy to explain why galaxies farther away travel faster than the ones closer to us. This book tries to prove the basic premise is wrong. If you change the term "is traveling" to "was traveling," dark energy disappears — it is no longer needed.

This is just one example of what I find wrong with today's theories in astronomical physics. Another example is "time" itself. No one has ever really defined or described what time is. How can you use it in theories without an absolute understanding of what it is in the first place? Science only has vague ideas or theories of what time really is.

If you ask ten physicists to define or explain what time is, it sounds like a class of third-graders explaining why the sky is blue. You end up with ten different ideas, none of which ever explain anything. In the end, they will eventually tell you they really don't understand what it is. Yet they talk about it as an absolute in their theories.

This book will try to explain time, and with that, explain or debunk most of the paradoxes, questions, and theories of modern physics that seems to baffle scientists today.

Since Albert Einstein, only one breakthrough has been made in practical science: Edmund Hubble discovered other galaxies. Until then the whole universe was the Milky Way. However, he also discovered they were flying away from us at great speeds. Other than that, we only have unprovable theories and conjecture. Even Einstein himself knew he wasn't quite right. He added the cosmological constant, and then he took it out; now dark-energy theory is trying to put it back in. He knew something wasn't right, he just didn't know what it was.

He also never embraced quantum physics because it did not fit in with his theory of time, or space-time as it were. Yet if you remove time from the equations, quantum physics and Einstein get along fine. But he was not going to remove time from anything for anyone.

Well in this book, I propose there is no such thing as time, in the scientific sense. Time is not real. Time is simply "the movie of reality," and the universe is a giant theater. Your perspective of time is related to your seat in the "cosmic theater." No one seat is the actual "real" seat. Time and reality are not the same thing. Let me explain.

How Einstein got the inspiration for his theory of time is he used what was called a thought experiment. As he was traveling on a train one day, he thought about what would happen if he was moving at the speed of light and looked back at the clock that was in the tower back in the center of town behind him. He surmised the hands of the clock would not move because the image of the clock would never catch up to him, so he would read the same time on the clock for as long as he kept moving at the speed of light. But the clock itself was not chasing him; it was the light bouncing off the clock that was chasing him. "Time," in essence, had stopped for him. But not for the people back in the town square under the clock. Time for them went on as usual; for them nothing changed. So what happened to Einstein traveling at the speed of light only happened to him. For the rest of the world, everything was normal. When he returned to the square, he, in reality, had not "travelled time"; he just missed out on what had happened while he was moving at the speed of light. It's not like the world went from A to F, so to speak. It's just he missed out on B, C, D, and E while he was moving at the speed of light. Now you can call this time travel, I guess, but I call it time manipulation. While he was gone, things everywhere else still happened at a regular pace, but Einstein is going to have to read about what happened in the papers or history books, depending on how long he was traveling at light speed. Again, it's not like it never happened because he wasn't there. He simply missed out on whatever happened.

This is just one of the alternate views I have to some of the theories in science today. Well let's get to it.

Chapter One
The Rules

In science there are basic rules that are accepted throughout the scientific world. These are the rules that we will be following throughout this book. I will point out where today's physicists bend them a little and sometimes completely ignore them. Here they are in no particular order:

1: *Light speed.* This is accepted by everyone, even me. 186,000 mi./s. Now whether that is a speed limit remains to be seen.

2: *E=Mc/2.* This rule is also accepted by everyone. It simply states energy is mass, and mass is energy — they are interchangeable. One is the same as the other. Keep this one in mind. We will be referring to it often.

3: *The conservation of information.* Don't get nervous; it's not as complicated as it sounds. Let me give you an example. You take a piece of paper and weigh it. Now take that piece of paper and burn it. If you capture all the smoke, keep all the ashes, and somehow weigh all that was left from that burnt piece of paper, you would come up with the same weight as it was before it was burnt. That whole piece of paper is still there. It's in a different form but all still there. In scientific talk, no information was lost. That's all there is to it, and this principle applies throughout the universe and throughout science.

4: *If it cannot be tested through scientific means, it is not science.* A hypothesis must be able to stand up to scientific testing. To be considered science, theories must be testable by some means, and these tests have to be repeatable. And results must be confirmed.

5: *Math is perfect.* Let's see about that. If you take three multiplied by three, you get nine. If you take nine and divide by three, you get three. If you take three and add three and add three, you get nine. No matter whether you multiply, divide, add, or subtract, the numbers are always going to be three or nine. So far so good.

Now try this. The way I heard this story is that Einstein proposed this problem to the Academy of Science. I am not sure how true that is, but wherever it came from, here it goes.

Three sailors go into a hotel. No, this is not a joke. They asked the manager for a room for the night. The manager tells them thirty dollars. So they each put in ten dollars and go up to the room. Soon after, the manager decides he charged them too much for the room. So he tells the bellboy to go up to their room and return five dollars to the sailors. On the way, the bellboy tells himself he can't give three guys five dollars evenly. So he knocks on the door, gives each sailor one dollar back, and keeps two for himself. Now each sailor paid nine dollars for the room. Three times nine is twenty-seven, and with the bellboy's two dollars, that's twenty-nine dollars. Where is the other dollar? So much for perfection. That's a flaw in math using only simple addition and subtraction. Imagine how many such flaws are in the seven blackboards of equations explaining concepts like string theory.

Another problem with the math in these long, complicated equations is it does not consider the transitions. Again let me give you an example. Let's say there are twenty birds on a telephone wire. Now take a shotgun and in one shot kill four them. How many birds are left? Math will tell you sixteen. But in reality, the other sixteen birds flew away when they heard the gun go off, so there are none left. What if the question is how many shots does it take to get rid of the twenty birds? Math would say five shots, but in reality, it only took one. As I said before, the other birds flew away when the gun went off. Again it is only a simple nuance but can change the outcome of the calculation significantly. So math is not so perfect after all. Math can give you an answer, but math is not *the* answer; there is a difference. Calculations alone do not prove a theory. There must be corroborative evidence.

6:   *If one part of a theory is proven wrong, you must throw out the entire theory.* This one is pretty much self-explanatory. And it has great consequences, as we will see later.

7:   *Newton's three laws of motion.*
   1.   A body at rest will stay at rest until acted upon by an outside force. Or a body in motion will stay in motion until acted upon by an outside force. Otherwise known as inertia.
   2.   Motion is produced when an outside force acts on an object. The heavier an object is, the more force it takes to move the object. Also this law describes friction. The more friction, the more force it will take to move the object.
   3.   For every action, there is an equal and opposite reaction, such as a rocket pushing flames out the back to make the rocket go forward.

Okay, those are the basic rules of science taught by professors in any remedial science class. In the upcoming chapters, I will demonstrate how science today will use one rule and simply ignore other rules or use part of the rule and completely disregard the rest of it. Are you ready? Here we go!

Chapter Two
The Time-Wedge Theory

As I said in the beginning, science seems to have more questions than answers. I believe what holds back most of today's scientific theories is time. Let me explain. Because of time, you end up with things like dark energy, string theory, and singularities in black holes, etc. Einstein's theories and quantum mechanics don't get along because of time. But if you remove time from the equation, the two get along just fine, and singularities are no longer a mystery. But how do you remove time? I believe Einstein was mistaken when he called time the fourth dimension and made it an integral part of every theory since. My theory removes time from the equations by separating time from reality. The easiest way to describe it is "time is the movie of reality." Now you must be asking, what do I mean by that? Well what is a movie? It is simply one picture after another shown at high speed to give the illusion of motion, just like the old kinetoscopes in the arcades. When you turn the handle, picture after picture would come up, and it looked like motion. But what is perceived as motion is not real. Another example is a cartoon, whereas you change the character's position on each cel little by little, picture after picture. So when you string them together, it looks like Bugs Bunny is walking down the street. Again the motion is not real. This is the basic concept of my theory.

When you look at something, you are not actually seeing the object you are looking at. You are seeing light rays bouncing off that object at 186,000 miles per second. So you are basically seeing one sheet of photons after another hitting your eyes instead of one picture after another, as in the cartoon analogy. You have a sheet of photons bouncing off that object and heading to your eyes, followed by another, followed by another, followed by another. Now light is moving so fast the delay here on Earth is undetectable to us. But when you get to cosmic scales, the delay becomes obvious, and it has ramifications that today's theories seem to ignore. Now this continuous stream of sheets of photons is the movie of reality, as I refer to it, and is what this book is based on. No matter what you are seeing, it is in the past. Am I making sense so far? Another little detail about light is that in complete darkness, you don't see anything. But that does not mean nothing is there. You just can't see it. So when you remove light, you remove time, as far as your eye's concerned.

Also anything that distorts the light along the way from its source to your eyes is nothing more than the special-effects department in a movie studio. Just because you saw a car jump over the Grand Canyon in a movie doesn't mean that a car really jumped the Grand Canyon. Another example of what you are seeing is not real is, in space you can look and see two identical galaxies on either side of the gravitational field. There are plenty of pictures of this phenomenon. But this does not mean there are really two identical galaxies. This is a special effect caused by gravity. The galaxies are not where we are seeing them. There are not even two galaxies. If you were to travel to where you are seeing them, you would be standing in space — there is nothing there. That galaxy you are seeing is somewhere behind the gravitational field. The light is being split by the gravitational field to give the illusion of two galaxies, one on either side. So to draw any conclusions from what you are seeing is completely nonsensical. What we are seeing is not real. I know this sounds a little weird so let's try another thought experiment to better explain it. Let's take Mars for example. It is approximately twenty minutes away at light speed. Now let's put a camera at MIT in Massachusetts, another camera on the moon, a camera on a satellite halfway between Earth and Mars, and another camera on the Martian moon Phobos. Now let's assume we have a quantum trigger so when you press one button, all four cameras take a picture at once. Now will all four pictures be identical? Of course not. All four pictures will be slightly different, depending on how far the light had to travel to reach the camera. So which one is the real picture of Mars? The answer is none of them. The only reality is what's happening on Mars itself. Each picture is real in relation to the distance the camera was from Mars. Also when the light leaves Mars and heads toward the cameras, Mars does not just stay there until the light reaches your eyes, or in this case the cameras. It is constantly moving in its orbit around the sun. So when you see the sheet of photons from Mars here on Earth, it is not a real representation of what is happening on Mars at that instant. Mars is twenty minutes farther along in its orbit, and it has also rotated a bit on its axis. And all the while, sheet after sheet of photons has been leaving Mars. But it will take about twenty minutes for you to see them. (See Figure 1). This is what I call the time wedge.

Now let's take this concept a little further. Let's say that there is a planet one half light-year away from Earth, and they have a telescope looking down at Earth. They don't see us as we really are. They see us as we were six months ago. It took our light that long to reach them. Now if it is summertime in the northern hemisphere here on Earth, they will see us with it being winter in the northern hemisphere. They see us on the other side of the sun, where we were six months ago when that sheet of photons left Earth. Now let's assume they have a "quantum leaper" on this planet and could get anywhere instantly. So they set the coordinates in their machine to what they were seeing in their telescope. But when they push the button and get to where they were seeing, they would be standing in space on the wrong side of the sun. For them to get it right, they would have to extrapolate Earth's orbit to predict where the Earth would be six months later than what they were seeing in their telescope. Again what they are seeing is not real. It's time.

Now take this theory to the edge of the visible universe 13.7 billion light-years away. The light left that galaxy 13.7 billion years ago, and remember, that galaxy has been traveling away from us at whatever speed for 13.7 billion years. So where is it today? How fast is it going today? What does it look like today? We have no idea. What we are seeing is the movie of that galaxy's life. What we see is that galaxy's baby pictures. To see how that galaxy grew up and how it turned out, well we're just going to have to wait 13.7 billion years. Now how big does that make the universe now? How big is the time wedge in this case?

My point is we have no idea how big the universe is today or where anything in it is today. We only know where it was, how fast it was going, and what it once looked like.

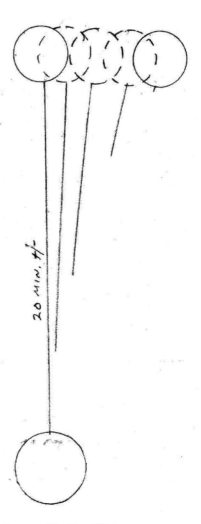

Figure 1: The Time Wedge
As the light travels to Earth, Mars is moving through space.

We could, however, figure it out with simple math, but nobody seems to be taking that approach. It's much more fashionable to come with exotic theories, some of which they themselves say can never be proven. Now remember the rules. If it cannot be tested or proven, it is not science. That's not my rule, it's one of theirs.

Now by looking at the universe in these terms, it answers many of the questions in today's physics. It gives science a way to deal with many of the paradoxes today's theories pose, such as the grandfather paradox in which you go back in time and kill your grandfather before he had your father. So how could you be born to go back and kill your grandfather?

Time is not real. Time is only what you see in respect to where you are in the universe. That's where Einstein had a blind spot. He insisted on using time in his theories without defining what time is and treating it as if time were a real thing, so much so that he made it a mainstay of his theories. That's why I say Einstein was just a bit wrong, and as I said before, he knew it too. He just didn't know why. So if you remove time, Einstein and quantum physics can finally get along.

For example, when you make a cell phone call, each electron emitted from your phone contains one bit of information. So if you are making a phone call and that stream of electrons says, "Hello, Frank," no matter where you are in the universe when you receive that call, it will always say, "Hello, Frank." Whereas in quantum physics, it predicts that phone call could say, "Good-bye, Mary," or "Your pizza is ready," or any number of things. But in reality, it will always say, "Hello, Frank." Now when you look at quantums in these terms, it makes more sense. I will get into this much deeper later on.

Another example of quanta versus reality is, on one of the shows, a professor from MIT stated if he walks down a set of stairs and takes a left, in some other universe he walks down the stairs and he takes a right. This is also wrong. The information encrypted in those photons leaving his body will always show him taking a left. If you are on the moon, Mars, another galaxy ten light-years away, or wherever, those sheets of photons are encrypted with information — nothing can change. You can warp it or bend it with gravity or distort it somehow, but he will always take a left.

Light can be manipulated, just as a circus mirror distorts light and makes your head look like it's two feet tall. This does not mean your head is too tall; it's the light just being manipulated by the circus mirror. Again what you see is not real. It is time.

So that is my time-wedge theory. Am I making any sense so far? Remember, I am not using some far-out data that I came up with on my own. I just use the examples science uses. I am just using the known facts in a completely different way and with a logical mind frame. In fact, I am using their facts to question their theories.

This is just one example of what I mean by science using one part of the theory and disregarding the other part. They use light or time as a real component of the theory, disregarding the time it took for the light to get here.

This is why I believe the time-wedge theory answers many of the questions posed by science today, and it may be the theory of everything that science has been looking for since Einstein. Time is not a line; it is a wedge. Also time is not real, so it cannot be used in any

calculations or as a component of any theory — or at least not the way it is being used today.

Chapter Three
Multiple Universes

Now if you take the time-wedge theory that this book proposes — that time is a movie of reality, and the universe is a giant theater — then wherever you are in the universe is a seat in the theater. That seat in the universal theater gives you your own perspective of the universe. Your view of the universe will not look exactly like anyone else's (see Figure 2). We have to get away from the notion that our perspective of the universe is the only one that counts and that we can look up and decide or describe what the universe is, was, and will become because, in reality, we have no idea what or where things in the universe are today. We can only look up and extrapolate the information we see to predict where everything in the universe may be or may look like now.

We cannot say the universe is this or that. We can only speculate by analyzing the data in the past tense. The math involved is simple enough. You just take the information our instruments give us and determine how fast and in which direction it is traveling and how far away it is. With this process, we can determine where it might be today and what it might look like today based on the sheet of photons that left that object however long ago. But even this would still be a guess because the gravity from other objects in space may have altered its trajectory along the way. Because of many other variables that could have affected the object, you will not really know where it is or what it looks like in

11

reality. For that you must wait until the corresponding time to distance has passed to know for sure where it is or what it looks like. For instance, if the object is one hundred light-years away, you will have to wait one hundred years to know where it really is today, which, in that case, you would be seeing one hundred years ago again and so on. Does that make any sense?

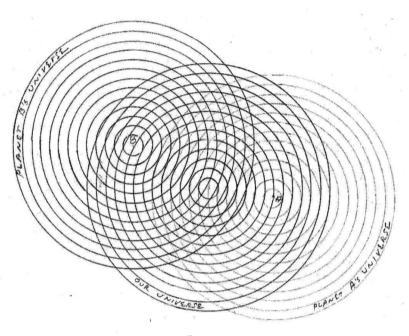

Figure 2

At each intersection the planets each see a galaxy at a different phase of its evolution which one is real? Does planet A have galaxies we cannot see or does planet A see the edge of the universe.

Let's do another thought experiment to see if we can clarify this. Say you are on a galaxy 150 or so light-years away from the Earth. You could look down on Washington, D.C., and watch President Lincoln get assassinated. Yet even if you had quantum travel and could get to Washington, D.C. instantly, you could still do nothing to stop it. You would be too late; it already happened because what you are seeing on your galaxy is a "movie" of Lincoln getting shot. Remember, it took 150 years for the photons to reach you. So again what you are seeing in your universe is not reality. Just like when we look into space we are not seeing reality — we are seeing a "movie" of what already happened throughout the universe.

This would also explain the grandfather paradox. You cannot go back in time and kill your grandfather. On what Earth would you go back to? Where exactly in the universe would it be? The image of your grandfather is in space somewhere. But that is all it is — an image, not reality. The Earth has been traveling with the solar system through the Milky Way since way before your grandfather had your father. It is no longer where it was when your grandfather was young. So exactly where do you go in "time" to kill your grandfather?

Here's another one. A physicist states that while standing in downtown New York, there are dinosaurs walking down the street with him right now. And he's right. You just have to be on a galaxy seventy million light-years away to see them. You could theoretically look down at Earth from there and see dinosaurs in New York. To you they would be real. You could tell your friend to look in your telescope and see these huge animals on this planet you found. You let him look in your telescope, and he will see the dinosaurs too. So on that galaxy you are on or in that universe that you are in, there are dinosaurs on Earth. But what you won't see is the physicist. For that you will have to wait seventy million years. That's how long the photons containing the physicist's image will take to reach you in your seat in the universal theater. Is this beginning to make sense?

Now this leads us to another conclusion: If there are an infinite amount of seats in the universal theater, that means there are an infinite amount of universes, just like the math predicts. This is an example of why I believe you don't need Star Trekian theories to explain multiple or parallel universes, just simple logic. Mr. Spock would have no trouble understanding multiple universes. Does that not make a little more sense than the explanations, or lack thereof, proposed by today's scientists, such as multiple universes just popping in and out of existence bumping into each other or black holes leading to other universes somewhere that no one can explain?

Today's science implies everything is happening everywhere at the same time, and with the right equipment or the right circumstances, you can jump around to any time or place in the universe you want. I say everything happens just once somewhere. And one sheet of photons containing that information goes into space at a time, not altogether everywhere, and once it has happened, that's it — it is etched in stone, so to say. You cannot go back and change it. They also claim that there are other dimensions or universes where up is down, right is left, good is bad, and so on. And in these other universes, they themselves exist with different traits than they have here on Earth. They say in some other universe they are taller or have red hair or whatever. Really? How about this? Say you are fifty years old; you can only exist in the universe up to fifty light-years away from Earth. That's as far as the light bouncing off you could travel in the time since you were born. This means in the universe beyond that you do not even exist, and in the universes you do exist in, you are exactly like you were when the light containing your photons left Earth. You're not taller, you're not shorter, and you're not meaner or anything else. You are you, no matter where in the universe someone sees you from. The only difference is where in the movie of your life, which is traveling through space one sheet of photons at a time, that person is seeing you. If they are observing Earth from ten light-years away, they would see you at forty years old. If they are twenty light-years away from Earth, they would see you at thirty years old, and so on. So what's the mystery? Whereas their theory is based on numbers and calculations containing time without really defining what time is. This is what I call a Star Trekian theory. They make great shows, but they make no sense scientifically when you really think about it.

Here is another example science likes to use to explain multiple universes. A scientist shines a flashlight into a glass apple; it sends out light in multiple directions around the room, and he uses this phenomenon as justification of multiple universes. But just because the light from the flashlight is being sent in different directions does not mean that there are multiple flashlights. There is still only one flashlight, and that is the source of all the other lights in the room. If you were to go to where the other lights are, there would not be a flashlight there. It is simply an effect of the light passing through the apple. The flashlight is real; the other lights are time. Then he goes even further. He states, "Light is a particle, and he is made up of particles. So just like there are other versions of the light, there must be other versions of him." There may be other lights in the room, but there is only one flashlight. There may be other images of him, but there is only one him. There is only one reality and an infinite amount of perspectives of that reality, which is what I call "time."

I could give you many more examples of wild ideas that are a byproduct of multiple-universe theories using time as a component as today's science does. But I do believe I have made a good argument for my theory, which is much simpler, and it does not lead to any paradoxes or supernatural outcomes. Well that's my explanation of multiple universes. What do you think?

Chapter Four

Dark Energy

Here is today's theory on dark energy. The farther away a galaxy is from Earth, the faster it is traveling away from us. Therefore, there must be some unexplained force in space pushing these galaxies away from each other faster and faster. The more space, the more force. Science calls this force dark energy. Makes sense, right?

Now try this. If you change the word in the theory "is" to "was," as I explained in the previous chapters, all of a sudden dark energy disappears. Let me explain.

When you are looking into space, you are looking back in time. This is a basic tenet of science. Yet science states that these faraway galaxies *are* moving away from us at whatever speed corresponding to its distance from us when the correct statement is these faraway galaxies *were* moving away from us at whatever speed. Science claims its data shows the galaxies are traveling away from us at a one-to-one ratio, meaning if a galaxy is a billion light-years away, it is traveling through space twice as fast as us. If it is two billion light-years away, it's traveling three times faster than us and so on and so on until you get fourteen billion light-years away, the edge of the visible universe. Now I have no doubts this is what the data shows. But their theory seems to disregard the fact that they are looking back in time. If you look at the same data using the time-wedge theory, which in essence changes the word "is" to "was," the universe stops accelerating, and in fact the data shows the universe is slowing down (see Figure 3).

But let's go with the dark-energy theory for a moment. Let's try another thought experiment. The galaxies are traveling away from us at whatever speed corresponding to their distance from us. We can see about fourteen billion light-years away. Now we have to assume everyone in the universe can see the same distance, fourteen billion light-years, no matter where in the universe they are. That means no matter which direction you look — up, down, left, right, whatever — you see fourteen billion light-years away.

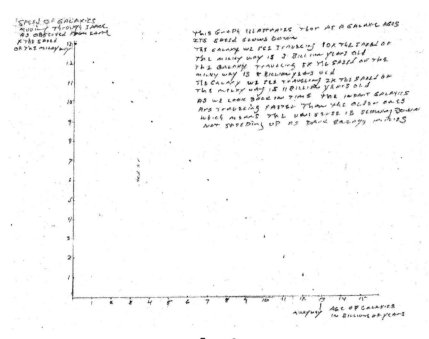

Figure 3

Now let's go to a galaxy seven billion light-years away from the Milky Way and let's bring all our equipment and detectors with us. What would we see? Well using the dark-energy theory as is explained, if we look forward in respect to the Milky Way, all the galaxies would still be moving away from us at a one-to-one ratio. No problem here. But what happens when we turn around and look back toward the Milky Way? Dark energy would imply we would see the galaxies behind us slowing down at a one-to-one ratio for seven billion light-years until we saw the Milky Way, and then beyond that, the galaxies would again begin accelerating for the next seven billion light-years at a one-to-one ratio until we saw fourteen billion light-years away from where we are now. And the galaxy that was fourteen billion light-years away from us would be traveling at the same speed as we are (see Figure 4). How does that work? If dark energy is universal, how can a galaxy fourteen billion light-years away be going the same speed as the one we are on. See what I mean? The basic premise of dark energy makes no sense, unless dark energy only emanates from the Milky Way and works in one direction.

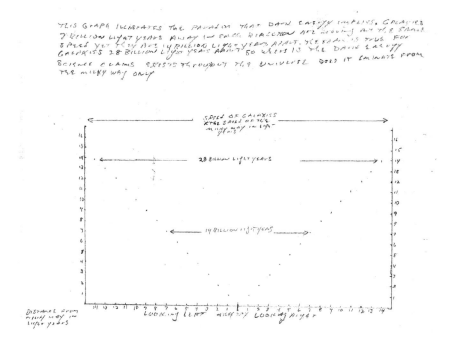

Figure 3

This is another example of proposing a theory and disregarding the implications of that theory to everything else. You can't just look at things one way, say it's a fact, and just move on. You have to take it to the next step and look at all the ramifications of the theory, as I just did. There are many more examples of conflict between theory and data. So it seems to me, when you take dark-energy theory a little further, it begins to fall apart.

Now let's look at the universe using the time-wedge theory. If a galaxy fourteen times farther away from the Milky Way was going fourteen times faster than the Milky Way, and a galaxy ten times farther away was going ten times faster and so on, this would imply the universe is slowing down, not speeding up. The older a galaxy is, the slower it is moving through space. (see Figure 3).

I believe if you were on that galaxy fourteen billion light-years away and looked back at the Milky Way, we would be traveling fourteen times faster than that galaxy. We would also be an infant galaxy to them. But then again, the Milky Way is only about 12½ billion years old, so they wouldn't even see us from that galaxy. As far as they are concerned, the Milky Way has not even formed yet. But you know what I am trying to say. This brings us back to the multiple-universe theory I proposed earlier. In their universe, the Milky Way does not even exist. Yet here we are.

This brings up another question. How many galaxies are there fourteen billion light-years away that are only twelve or thirteen billion years old? We would not be able to see them because their light has not reached us yet. No one knows how many there are. Is it a hundred, a thousand, a million, or probably billions? Now how can you calculate the mass of the universe when you are not even sure how many galaxies are out there?

Another paradox of dark energy is if it works in all directions, shouldn't a galaxy fourteen billion light-years away from us be pushing us fourteen times faster than it is going or does dark energy only work in one direction, away from the Milky Way? Are we the source of dark energy, as the theory seems to imply?

Another paradox of dark energy is the largest structure in the universe. It is a cluster of galaxies that stretches five billion light-years across — that's more than one-third the size of the visible universe. Where is dark energy in this case? If dark-energy theory is right, a structure like that should not even exist. Using today's theory, the galaxy at one end of the structure should be pushing the galaxy at the other end of the structure five times faster than it is going. Yet the structure seems to be gathering up other galaxies with its gravity, supposedly the weakest of the forces. But we'll get to that later. The whole idea of dark energy seems to fall apart when you look at the universe and not just pick out one observation that seems to prove your point and ignore the facts that seem to conflict with your theory.

Here is another thought experiment. From our place here in the universe, we can see fourteen billion light-years to the left and fourteen billion light-years to the right. We can assume from the Hubble deep sky photograph that there are galaxies at both extremes. That puts those galaxies twenty-eight billion light-years away from each other. So if dark energy is real, shouldn't those galaxies be pushing each other away twenty-eight times faster than they are going, or the Milky Way is traveling through space, or whatever benchmark you wish to use. The number should be twenty-eight times something. Yet when we observe them from Earth, they're both going the same speed. Again this implies dark energy emanates from the Milky Way and works in one direction. The theory implies the Milky Way is pushing the other galaxies away, but the other galaxies don't seem to be pushing us away. Do you see what I mean? The whole theory is chaotic and full of paradoxes, yet today's science swears by it. And while a new area of research was formed, now some thirty years later or so, not one explanation or idea has been put forth to explain dark energy — just endless research looking for something they themselves say they have no idea what it is, how it works, or where it comes from. And by the way, a Nobel Prize was awarded for this theory. Well that is my argument against dark energy. What do you think?

Chapter Five

Gravity

O kay, here we go. Newton described gravity as a string of something hold-ing objects together. He just couldn't quite describe what the string was made of. But his math is absolute; it is still being used today to launch satellites to a specific orbit, calculate trajectories, and for all other calculations related to the orbit of objects in the universe. It was Newton's calculations that enabled the Apollo 13 capsule to return to Earth after a catastrophic explosion disabled the spaceship. Without Newton's calculations, the astronauts would've been lost forever. But by using Newton's math, they determined how to use the moon's gravity to slingshot the capsule back to Earth.

Yet because Mercury's orbit does not quite fit in with Newton's theory, its orbit being a bit irregular around the sun, Einstein threw Newton's entire theory out, as if there is no other way of explaining Mercury's flower-petal-shaped orbit, such as its massive density, magnetic fields, proximity to the sun, or any number of other factors. But back to the rules: If one flaw is in a theory, the whole theory must be thrown out. Well rules are rules.

But back to the beginning. Newton described gravity as some kind of string holding objects together by pulling on each other. But he could not quite describe what this string or these strings were made of. But the math de-scribing these strings pulling on objects is perfect, except for Mercury of course. It not only works throughout the universe, it also works on Earth. Remember, it was supposedly a falling apple that gave him the key to formulating his idea of gravity. As I said before, it is still being used today.

Einstein's description is quite different. He believed space and time were woven together into some universal fabric he called space-time. And the mass of objects in space were causing a dent in this fabric of space and time. The size of the dent in space-time caused by the object is directly correspondent to its mass; the larger the mass of the object, the bigger the dent in space-time, the bigger the dent in space-time, the more gravity. In the case of multiple objects of different masses in close proximity of each other, the smaller object will fall into the dent in space-time caused by the larger object. So in this case, science states the smaller object is being "pushed" by space-time, not "pulled" by gravity toward the mass of the larger object. Or another way to look at it is, space-time is pushing the apple to Earth and not the Earth is pulling the apple to the ground. Now as you see, the difference in the two theories is significant. But because of the flaw of Mercury's orbit, Einstein's theory won. But as you will soon see, Einstein's theory also has a few flaws in it. Now the only proof of Einstein's space-time theory is a picture of a total eclipse in which the light from some of the stars behind the sun that were not supposed to be visible from Earth was curved or bent around the sun to be photographed here on Earth, therefore confirming the fact space-time was curved or bent by the mass of the sun. And this curve in space-time is what gravity is. End of story, right?

Well you can move or bend light several ways. Magnetism moves light. Science uses powerful magnets to keep super-hot plasma from burning through equipment during experiments. We have all seen heat waves bend light while looking across a paved surface in the summertime heat. Any number of energy waves can affect light. Here on Earth, we see the sunrise about two minutes before it actually rises above the horizon. This is caused by the atmosphere bending the sun's rays over the horizon. Yet in the sun's case, science tells us it is the dent in space caused by the sun that allows us to see the stars behind it. Well I believe it's the sun's atmosphere. Scientists tell us on Earth it's our atmosphere, but in the case of the sun, it's a dent in space-time. Does that seem logical?

But let's take a look at the dent-in-space theory. Let's follow Einstein and see where he leads us. The example science often uses to describe Einstein's theory uses a trampoline, a bowling ball, and some marbles. So let's do another thought experiment using this example. We place a bowling ball in the center of the trampoline. Now this causes a dent in the middle of the trampoline. The bigger the ball, the bigger the dent in so-called space-time will be. Now let's send the marbles around the bowling ball, all at different speeds, to simulate the planets. This example works except for one thing: The planets or marbles form a conical shape. You have a cone-shaped solar system. Yet the planets orbit around the sun on a flat plane. The trampoline example does not match what you see with your own eyes on any of those shows about space. The theory does not match the data (see Figure 5). Another point on this: Galaxies are flat. If Einstein is right, wouldn't galaxies be conical shaped due to the supermassive black hole in the center of the galaxy bending the fabric of space-time almost to breaking point, as the theory of black holes implies? Shouldn't the universe be full of cones? Again the theory doesn't match the data.

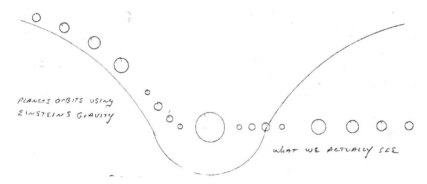

Figure 5: Distortion of Space Time as described by Einstein

Flaw number one: Einstein's dent in space-time theory states all objects travel in a straight line, and only when it encounters the curvature of space-time caused by a massive object will its path be altered. The object is then supposedly being held in orbit by space-time pushing against it, similar to a car door pushing against you as you go around a sharp curve. Now this would imply friction between the fabric of space-time and the object, wouldn't it? Then logically you would expect the rotation of the object to eventually come to a one-to-one ratio with its diameter, meaning the circumference of an object will equal the distance it travels in its orbit around the mass. For instance, if the marble on the trampoline has a diameter of two inches, it travels two inches per rotation along the fabric of the trampoline. The laws of physics will not allow the marble to spin slower or faster. Now look up in the sky; throughout the universe there are examples of planets spinning slower than physics allows and some examples of celestial objects rotating much faster than physics allows. For example, the moon only rotates once in its orbit around the Earth. So it takes about twenty-eight days for the moon to turn once. Shouldn't the friction of the moon against the fabric of space-time be speeding up its rotation. Another example is Jupiter. It rotates once about every ten hours. Shouldn't friction be slowing down its rotation. There are even examples such as Uranus, which spins in the opposite direction of the other planets and is tipped over on its side and is essentially rolling through space. How can this be? The law of friction has to be ignored to make Einstein's theory work.

Flaw number two: If space-time is pushing the Earth to hold it in its orbit around the sun, explain to me high tide. Shouldn't space-time be flattening the back side of the Earth in respect to the sun? Yet if you ask a scientist what causes high tide, he or she will tell you the moon does this by pulling on the surface of the ocean. The trampoline example does not even begin to explain this. The dent in space-time theory would predict a flat spot on the part of the Earth in contact with space-time. But what we have instead is a bulge on the back side, and not only that, we have a bulge on the side facing the moon. It is also a fact that the Earth is pulling on the face of the moon, causing it to bulge approximately seven meters on the side facing the Earth. The space-time theory cannot explain these facts. Newton's theory of a string of some sort between two objects explains this perfectly.

Flaw number three: Now let's look at Jupiter's moon Io. Satellite images show it to be extremely volcanic. The way science explains this is that Jupiter's gravity is stretching and pulling on the moon surface as it rotates and goes around the planet. This constant stretching causes friction within the moon, which in turn generates heat and causes the volcanic activity. Makes sense. But how does a dent in space-time cause Io to stretch and contract. Also when comet Shoemaker-Levy 9 passed by Jupiter, the planet's intense gravity tore it apart and stretched it into what was described as a string of pearls. Stretching is a product of pulling. How does a dent in space-time explain this? Newton's theory does.

For Einstein's theory to be correct, the Earth's oceans would have to be falling into the dent in space-time caused by the moon. Give me the math on that. Also the face of the moon would have to be falling into the dent in space-time caused by the Earth.

Flaw number four: Science will also tell you the earth's gravity is pushing on us, holding us to the earth. Then what happens when I jump up? Do I bend the fabric of space-time and it pushes me back down? Or do I pierce one fabric, and there is another fabric behind it that pushes me back, and the other fabric seals back up or something? When a plane takes off, does it stretch space-time up to 35,000 feet, then ride along the fabric until it reaches its destination? Then I assume the fabric pushes the plane back to earth when it lands. Exactly how does that work? Does someone who is six-feet-four have a different fabric than a child three-foot-tall does? The same fabric that holds the moon in orbit also holds the tall guy and the toddler to the earth. Again exactly how does that work? And how many fabrics are there?

Why does a rock sink to the bottom of the ocean? Is there space-time underwater? If it's the fabric of space that holds us down to the earth, shouldn't the rock be pushed below the surface of the water and stop there? What pushes it to the bottom of the ocean? Newton's math works on the earth, underwater, in a cave, and throughout the universe. A dent in space-time has no answer for this.

Flaw number five: Again I ask how many fabrics are there, and on what plane do they lie on? Exactly where is up? In the trampoline analogy, there is one sun, one dent, in one fabric, yet we have satellites in many different orbits around the Earth. Some are relatively close to Earth, such as the International Space Station. Others, like stationary satellites as they are called because they stay in the same place in the sky in relationship to the stars, are thousands of miles above the Earth. Do they each have their own fabric? What about satellites in circumpolar orbits, which means instead of orbiting around the equator, they orbit over the North and South Pole as the Earth rotates beneath it? Do they have their own fabric, or are they pushing against the same fabric as the other satellites? How can the Earth cause a dent in one fabric in two directions? How about when a satellite runs out of fuel and descends to Earth? Does the fabric follow the satellite in until it crashes into the earth? Or does the satellite go from fabric to fabric to fabric until it crashes? Exactly what are the mechanics of this process? How does it work? Einstein's theory seems to be lacking a few details.

Flaw number six: Throughout the universe, there are binary star systems where one of the stars has exploded and what remains is a black hole. There is plenty of proof of this. There is also proof that the black hole is consuming its sister star by pulling the matter away from it. Yet the sister star stays at the same orbital distance away from the black hole. It is just the matter that is being sucked in. Shouldn't the dent in space-time be "pushing" the whole star toward the black hole? How does the dent in space-time "push" just the gases from the star into the black hole yet leave the star at relatively the same distance?

Flaw number seven: As our equipment gets more and more sophisticated, science has been finding other planetary systems throughout the galaxy. One of the ways they find them is by detecting a wobble in the parent star. The gravity of the planet pulls the star back and forth a bit as it orbits around it. This phenomenon is described as the center-of-gravity effect, which Newton's theory explains perfectly. To have a center of gravity in the trampoline scenario, the bowling ball would have to be falling slightly into the dent in space-time caused by the marble. In essence the marble would be causing the bowling ball to move. This does not happen. Just the term "center of gravity" debunks Einstein. The center of gravity is the point between two objects where the pull of gravity is equal. Gravity between two objects in the trampoline scenario would imply they are both falling into a dent in space between the two objects. The larger object would be closer to the center dent, and the smaller object would be farther away. But what object is in the center causing the dent that the two objects are falling into in the first place? This is getting a little out of hand, isn't it? Let's use a real-world example of center of gravity. In space there are what are called LaGrange points, where an object in space could stay in a specific place in space without any propulsion or external corrections necessary. They are points between the Earth and the moon, the Earth and the sun, or whatever. There is a sweet spot between two celestial objects where the gravitational pull, not push, is equal, so the object is not being pulled toward any one object. For Einstein to be right, the objects at the LaGrange point would be balancing on the crest of two dents, one caused by the moon and one caused by the Earth. For Newton to be right, the object at the LaGrange point is being pulled equally by the Earth and the moon. Which scenario seems to make more sense to you? Again I have to ask, exactly how does this work? A dent in space-time cannot explain the "center of gravity."

Flaw number eight: Let's go back to Apollo 13. Remember, they were heading off into deep space in a damaged vehicle. The only way to get back to

Earth was to use the gravity of the moon to slingshot them back toward the Earth. Einstein's theory, that the moon causes a dent in space-time, works if you are looking down at the trampoline and the bowling ball from above. When you roll a marble past the bowling ball, as it enters the dent, it will curve. And if you roll it by at the right speed and the right trajectory, you could make the marble go right around the bowling ball and come back to you. But we are on Earth. The gravity on Earth keeps the marble in contact with the trampoline as it comes out of the dent. Now if you took the experiment into space to a zero gravity environment, as the marble goes around the bowling ball, it also goes into a dent. So as the marble leaves the dent, it is not only coming back toward you, but it is also rising out of the dent. Without gravity to hold it to the trampoline, it does not stay on the same plane. So the marble flies out of the dent and up into space and completely misses you. It comes toward you, but the marble, without gravity, would fly over your head. It is no longer on the same plane as the Earth. If NASA used Einstein's gravity, Apollo 13 would've never made it. It would have been thrown, eventually, out of the plane of the Milky Way (see Figure 6).

Flaw number nine: On the earth, there are variations in the gravitational field all around the globe. The biggest example of this is in and around Hudson Bay in North America, gravity seems to be lacking. Scientists tell us that this is because of glacial formation during the last Ice Age. The ice was over two miles thick, and the tremendous weight of all that ice compressed the earth's crust. This pushed the earth's crust to each side, leaving less mass where the glacier was. It seems in this case too, less mass, less gravity. How does a dent in space-time explain this? Does the fabric of space-time have variations? Is space-time heavier here than over there? Also the moon has one-sixth the gravitational force of the earth. So does the fabric of space-time weigh less around the moon than it does around the earth? I understand the moon's dent isn't as deep as the Earth's dent, but how does the depth of the dent equate to the force of the fabric? And how about Jupiter, where a man could not even walk around, assuming there was a place to walk, because he would weigh so much? How does space push harder on Jupiter than it does on Earth? How does the depth of dent in space correlate to the fabric of space-time getting heavier and pushing harder?

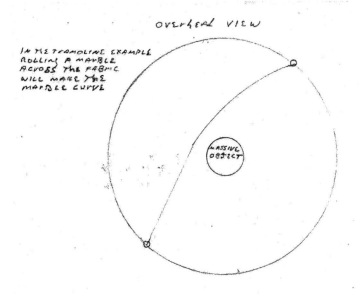

OVERHEAD VIEW

IN THE TRAMOLINE EXAMPLE
ROLLING A MARBLE
ACROSS THE FABRIC
WILL MAKE THE
MARBLE CURVE

MASSIVE OBJECT

BUT IN SPACE THERE IS NO GRAVITY
SO THE OBJECT WILL CURVE BUT IT WILL
ALSO CHANGE TRAGETORY AND
BE SHOT INTO A DIFFERENT PLANE
IF EINSTEIN IS RIGHT APPOLO 13 WOULD
HAVE BEEN SHOT OVER THE EARTH

SIDE VIEW

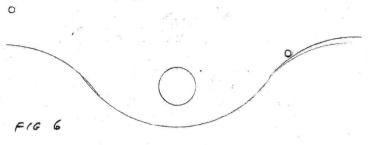

FIG 6

Figure 6

30

Flaw number ten: For millions of years, the moon has been receding from Earth. The moon used to be much closer than it is now. Also the Earth's rotation used to be much faster. Some say the Earth's day was only six hours long at one time in its history. So what happened? Well first, the moon is traveling so fast that the Earth's gravity isn't quite strong enough to hold onto it completely. So little by little, it gets farther and farther away from the Earth. And this is still happening today. Each year the moon gets about 1½ inches farther away. So now the trampoline theory would imply that the moon has also been slowly rising above the plane of the Earth as it climbs out of the dent in space-time. But there's no evidence of this. It seems it's been on the same plane for millions of years, whether the moon is 25,000 miles away or 250,000 miles away.

Science also states that the moon's gravity is what's causing the Earth's rotation to slow down. As the Earth rotates, the pull of the moon causes a bulge in the oceans, which again is high tide. As the Earth turns, the pull from the moon on these bulges in the ocean, just before it turns away from the moon, causes "drag" on the Earth, which in turn causes it to slow down just a bit. As little as this is though, over millions of years it caused the Earth's rotation to slow down to get to where we are today, about twenty-four hours in a day. Without a physical interaction between the Earth and the moon, this cannot happen. There has to be direct contact between the two bodies. A dent in space-time cannot begin to explain this.

Flaw number eleven: Now let's take a look at stars in general. A star is a nuclear reactor. What stops it from just exploding from the force of the nuclear fusion is gravity. The force of gravity is equal to the outward pressure of the nuclear reaction going on inside. How does a dent in space-time hold the whole star together? A dent in space could presumably hold the sides of the star in, but what about the top? How does a dent in space form a sphere? And what about a supernova? Just before a star goes supernova, the core collapses — not the whole star, just the core. Again I am not making up phony scenarios to prove a point. All these are real questions based on what accepted science is. Does the core have its own fabric of space-time?

Flaw number twelve: What would a comet's orbit look like in the trampoline scenario? Everyone knows a comet's orbit is elongated. But shouldn't it also have an arc to it, as it goes in and out of the dent around the sun and reaches the outer edges of the solar system? The trampoline theory implies that there should be an arc somewhere in the orbit. How does it stay on a flat plane and follow a dent in the fabric of space-time? Again the data doesn't match the theory.

Flaw number thirteen: Now it's time for gravitational thought experiment. Science claims if you have a teaspoon of neutron star material, it would be so dense and weigh so much it would fall right through the Earth. So let's try this. Take that teaspoon of neutron star material and place it in space next to the Earth. What would happen? Would the piece of neutron star material go flying toward the Earth and crash into the Earth? Or would the Earth be pulled toward the teaspoon of neutron star material?

Using Einstein's theory, the Earth would fall into the dent in space-time caused by the neutron star material. Newton's theory implies the neutron star material would pull the Earth to it. Now what happens when the Earth and the neutron star material meet? Do they just spin around each other? Well if you ask me, the Earth should get sucked right around the neutron star material, similar to a dot being eaten by a Pac-Man. The result is they would reach their center of gravity. The dot does not move, the Pac-Man moves, and that teaspoon of neutron star material would be embedded in the center of the Earth, greatly increasing the Earth's mass, whereas its size would only increase by a teaspoon. Either way a dent in space-time has no explanation.

Fundamentally if Einstein is right, then gravity is not even a basic force. It is an effect; it requires a fabric and a mass. The interaction of the two results in gravity. Therefore, by definition gravity cannot be a basic force at all. See what I mean about paradoxes in science and theories that ignore facts that get in the way? Science claims there was no space or time before the Big Bang, just energy. Well what was holding the primordial energy together if the four basic forces didn't even exist yet? What kind of energy was it if the nuclear forces didn't yet exist? Remember, the four basic forces were created in the first instant of the Big Bang. That is what science has been teaching for decades.

By now you are probably asking, how do I explain gravity? Well let me give it a shot. Science calls the four basic forces the strong nuclear force, the weak nuclear force, electromagnetism, and gravity. I call them the strong and weak nuclear force and the strong and weak magnetic force. Doesn't that make more sense than a strong and weak of one force and two totally independent forces where, in essence, you end up with three forces? Again gravity needing two things is not even a basic force if Einstein is right. My theory gives you two forces, a strong and weak of each. One of the magnetic forces is electromagnetism. The other is what I call gravitational magnetism. The weak magnetic force would be electromagnetism. It has a limited range and can easily be manipulated by outside forces. An example is a compass needle being affected by the giant iron mass of the ship itself, in which case the compass is enclosed in a binnacle with two steel balls on either side of it to counteract the interference from the iron of the ship itself. Whereas gravitational magnetism is affected by only one thing: mass. The more mass, the more gravitational magnetism, plain and simple. Gravitational magnetism is also cosmic. Everything in the universe is gravitationally bound to everything else in the universe at some level. Also gravitational magnetism is much harder to manipulate although I believe it can be done, as I will explain as we go on.

Let's use the sun as an example. It has the nuclear forces, it has electromagnetism, and it has gravity. Now how far do these forces interact, let's say, in our own solar system? Does the nuclear force from the sun reach the Kuiper Belt? Maybe, but not with very much intensity. It is extremely cold that far away from the sun. Does its magnetic force reach the Kuiper Belt? Again maybe, but not with much intensity. Yet the sun's gravitational force holds the Kuiper Belt objects in orbit. And not only that, it's also holding the Oort cloud in orbit around the solar system. Now the Oort cloud is so far away we can't even see it. It is hypothetical; science tells us that the Oort cloud is where comets come from. Supposedly if there is a disturbance in the Oort cloud due to some gravitational influx from whatever, the objects in the cloud will get tossed around. Some objects get ejected into deep space, but some will get thrown toward the inner solar system. The sun's gravity then pulls the comets from billions of miles away, yet science claims gravity is the weakest force. Gravity is strong enough to affect things the other forces barely reach.

The sun is also gravitationally interactive with Alpha Centauri. That's four light-years away. Is the sun nuclear bound to Alpha Centauri? Is the sun magnetically bound to Alpha Centauri? No, but it is gravitationally bound. Why does science ignore all the evidence that contradicts their hypothesis? But to explain the notion of gravity being the weakest of the forces, the example science commonly uses is to take a tiny magnet and pick up a paper clip, which in turn leads to the question, how can the tiny magnet pick up a paper clip with the gravity of the whole world pulling against it? Just the question itself debunks Einstein because Einstein claims gravity pushes, not pulls, remember? But back to the question. The obvious answer is magnetism is stronger than gravity. So by showing a tiny magnet is stronger than the whole earth, the conclusion is a little bit of magnetism is stronger than a whole lot of gravity.

One fact being forgotten here is gravity is proportional to mass. For gravity to interact, there has to be enough mass to interact with. If that magnet is so strong, try to pick up a Buick with it. All of a sudden, gravity seems a little stronger. But let's keep going with the paper clip. If you go to a scrap yard, they have these giant electromagnets attached to cranes that can pick up tons of steel very easily and move it wherever they want — and for the record, it can pick up a Buick too. Now take that same paper clip and touch it to the electromagnet. You will find you can easily remove it from the magnet. Does that mean you are stronger than the electromagnet? Of course not. It's just the paper clip; it does not have enough mass to hold it to the electromagnet. Their whole example is flawed. It ignores the laws of proportion in relation to force.

Back to gravitational — or basic magnetism, as I like to call it. Unlike electromagnetism, which is directional, having a north and a south pole, basic or gravitational magnetism is omnidirectional — it has only one central pole. Its force is equal in all directions. A good example of this was an experiment performed a few years ago on the International Space Station. An astronaut put sugar and salt into a plastic bag and shook it up. To everyone's amazement, the grains did not just float around in space, as you would think would happen. They began to clump together in a process known to science as accretion. To me this is gravitational or basic magnetism in action on a small scale. No dent in space-time required. The grains were simply attracted to each other like little spherical magnets. They stuck to each other top, bottom, left, right — all directions. This is proof of the basis of my theory. Every object, no matter what size it is, has gravity. Gravity is a basic property of matter, and I believe this goes right to the atomic scale.

When science describes planetary formation, it claims the dust that remained around the sun after it formed began clumping together through the process of accretion. Then when the object accumulated enough mass, gravity took over, and ultimately a planet was formed, and this is also how everything in the universe got started, from the smallest asteroid to the largest star. But what drew the dust together in the first place? Why didn't they just drift by each other? What made them stick to each other in the beginning? Why do the salt and sugar crystals clump together and not just float by each other in space? Science gives no explanation for the first stage of planetary formation. Does it just happen? Nothing in the universe just happens. Gravitational magnetism explains it. The math is there, but again I'm not going to get into that in this book. From the micro to the macro, basic gravitational magnetism works, and until science comes up with something better than a dent in space-time, I'm going to stick with it.

There is a manufacturer that has come up with a desktop toy that inspired this next thought experiment. They are spherical magnets that stick together in any direction. You can build squares, balls, pyramids, or whatever you want with these little, round magnets. Now by using these magnets, you can much better demonstrate gravitational magnetism on a flat plane with no dent in space required, and the analogy actually matches what you see throughout the universe.

In my example, you place a large, spherical magnet in the center to represent the sun and eight smaller ones of different sizes to represent the planets going around the sun. With the magnets traveling at different speeds and at different distances, you can demonstrate Newton's gravity perfectly. The magnets act as if a string is holding them together or, more specifically, a rubber band. Now if the speed of the small magnet is too slow, the large magnet sucks it up, and it ends up crashing into the sun, as it were. If the speed of the small magnet is too fast, the hypothetical rubber band breaks, and the smaller magnet takes off into outer space. Now different-sized magnets having different strengths due to size; traveling at different distances from the large magnet requires different speeds to stay in a stable orbit. It's all about proportion or mass as it is related to the speed and distance required to obtain a stable orbit around the sun. The spherical magnets are the perfect example of Newton's gravity. They work just like the planets around the sun, just like the math shows it should. Now this brings us back to what is the rubber band made of? Well gravitons of course. The more mass, the more gravitons. The closer an object is to another, the more the gravitons interact with each other, and gravitons exist in the subatomic scale just like photons and electrons, which should please the quantum people. Seems simple to me.

Science poses its own thought experiment to further discount Newton's theory. If the sun were to disappear, what would happen to the Earth? Would it fly off into space immediately, or would it take nine minutes before the Earth felt the effect of no gravity because nothing can travel faster than light? Would it take nine minutes for the force of the dent in space to rebound and be felt here on Earth? Newton's theory implies if you remove the sun — or in Newton's example, let go of one end of the string — the Earth flies off immediately. No delay necessary, no laws have to be broken. In the Einstein theory, the sun is removed, and then the dent in space begins to rebound, and nine minutes later, the Earth stops falling into a dent in space and flies off. I have got to see the math on that.

No matter how hard today's science and its explanations try to justify Einstein's theory of gravity, it just doesn't work, and the orbit of one planet and one picture of an eclipse is not enough evidence to throw out Newton's whole theory. I have just given you at least a dozen flaws in Einstein's theory. I know it is blasphemy to go against Einstein, but the facts are what they are, and I believe I just gave gravity back to Newton. Sorry, Albert. I believe I have proven my case. What do you think? At least I have given you something to think about.

Chapter Six
Dark Matter

The need for dark matter comes from two sources: one observation and one computer model. First let's look at the observation.

Back in the 1960s, an astronomer was observing the motion of stars in the Andromeda galaxy. Her observations showed that the stars at the edge of the galaxy were moving at the same speed as the ones closest to the galactic center. This observation seems contrary to what physics would predict. One would expect the outer stars to be traveling much slower than the stars at the inner part of the galaxy, similar to the planets orbiting the sun. Now this conclusion would be fine if galaxies and solar systems had formed under the same conditions, but they didn't. And I will get to that later. This phenomenon was, and still is, puzzling to astronomers. It seems there was not enough mass in the visible galaxy to make the stars move this way. There was simply not enough gravity, and the math showed there needed to be five times as much matter to achieve this result. So to explain what was happening, they came up with dark matter. The basic tenet of this theory is there is not enough visible matter to have enough gravity to explain the motion of the stars. So there must be another kind of matter that the gravity comes from, and with this conclusion, dark matter was born. The theory proposes that galaxies must be encased in a cloud of dark matter that is causing the whole galaxy to spin as a unit.

Now because we cannot see or detect this matter in any way, they call it dark, and after fifty years of trying to detect or explain it, we are no closer today than we were back then.

Now the computer model. We have entered the age of supercomputers. So to further understand dark matter, a scientist inserted a cloud of matter into his supercomputer to see what would happen, and again matter alone did not form a galaxy. So he added more and more gravity to the simulation until he got a galaxy to form. As it turns out, it took five times more gravity than the matter contained in the simulation to achieve this result, which further confirmed the need for dark matter, and this is where science stands today.

As science is trying to explain dark matter, they are using abandoned mines miles deep in the earth's crust where the sole purpose is to detect WIMPS (Weakly Interactive Massive Particles), which are one of the subatomic particles science believes is an explanation of dark matter. Now they have been running this experiment for over ten years all over the world, and while science predicts millions of WIMPS passing through the Earth every second, so far only two "maybe" WIMPS have been detected. We also have other experiments all around the world looking for dark matter, none of which have come up with the confirmation of any dark matter particle in any way, shape, or form. We also have satellites in space trying to detect dark matter particles, yet with all these resources and all this technology, science has come up with nothing. No one has any idea or any explanation of what dark matter is.

Again let me give it a shot. To explain dark matter, you have to go back to how the galaxies were formed in the first place — all the way back to the Big Bang. After the Big Bang, the matter that the galaxies were formed from was moving at least at the speed of light, and I believe it was even much faster than that, but we will get to that later. It was not a stagnant, primordial cloud of matter, as the computer simulation implies. This matter was moving at tremendous speeds when the matter began clumping together to form stars, which in turn formed galaxies. Now these clumps of matter moving forward through this primordial soup would be better explained with fluid dynamics rather than basic matter stagnant in a vacuum — for example, a river flowing over a boulder in the riverbed. As the water flows by the boulder, it creates an eddy or vortex in the fluid. Then the river's forward motion causes the water in the outer part of the eddy to spin at the same speed as the water at the center, much like we see in the observation of stars in galaxies.

For the evidence of the computer simulation to be relevant, you would have to believe that after the Big Bang, there was a cloud of stagnant gas or matter just sitting there. Then dark matter formed the galaxies. But then what? They all took off at the speed of light. The basis of the theory implies there was a big bang, then a stagnant cloud of gas, then galaxy formation, and then acceleration. Does that make sense? We all know galaxies are traveling forward at great speeds, even today fourteen billion years after the Big Bang. So common sense would tell you they had to be traveling forward even faster than they are today during their formation. And that's where I believe the answer to dark matter is. Back to Einstein's theory, energy is matter. The energy of their forward motion through the primordial soup, I believe, is the explanation of the missing mass, not some dark matter no one can find and no one can explain.

To do an accurate computer model, you not only have to put a clump of matter into your supercomputer, you have to have that huge cloud of matter moving through or with all the other matter in the primordial soup traveling at least at the speed of light, and as I said before, I believe it was much faster than that. Again the experiment or explanation doesn't seem to include all conditions or all the laws of physics that science claims were present during galaxy formation. They seem to be picking and choosing their conditions for the scenario. One scientist would tell you everything was moving much faster than the speed of light, at least for an instant, as the theory of cosmic inflation implies. I am not a fan of the theory, and I have an alternate explanation that I will get to later. But in the computer model, this is ignored. You can't have it both ways — either it was stagnant, or it was moving faster than the speed of light. Now matter moving at the speed of light contains an enormous amount of energy, and back to Einstein, energy is mass, and mass is gravity. The missing gravity is the energy of the forward motion. So my theory proposes that a galaxy is basically a Frisbee being thrown through the universe. Galaxies were only formed during a distinct period of time after the Big Bang during extreme turmoil. Solar systems are still being formed today. Galaxies formed under tremendous forward motion; solar systems formed in a stagnant cloud. Do you see the difference? Science doesn't seem to.

Now another example I like to use to explain the missing mass — or dark matter, as it were — is the BB example. No this is not an acronym for some crazy outlandish theory; I'm talking about a BB from a BB gun. Now again this is proven accepted science.

It goes like this. If I take a BB and throw it at you, it would just bounce off you and fall to the ground. Now if I take that BB and put it in air rifle and shoot it at you, it will go through your shirt and pierce your skin to whatever degree. Now if I take that same BB and put it in the super gun that they have at one of these universities to do such experiments, this gun shoots that BB at 30,000 feet per second; it will destroy a bowling ball. This is documented; it is not a hypothetical. Einstein's speed-limit theory implies that as you increase the speed, you increase the mass. Now I have no problem with this, but the science implies the BB actually gets bigger and bigger the faster it goes till it gets so big and dense that there's not enough energy in the universe to move it to light speed. My theory proposes that the speed alone accounts for the extra mass, and the experiments seem to prove it. The BB is always the same size. It does not grow, and it does not get bigger, yet it contains a considerably more amount of energy, and it causes increasingly more amounts of damage. This energy caused by the increased acceleration is the missing mass everyone is looking for.

Now by looking at energy or mass in these terms, you can explain the missing mass or the movement of the stars in the Andromeda galaxy. The whole universe was moving at tremendous speeds during the formation of galaxies, and this forward motion is the missing mass that causes galaxies to behave as they do. Now solar systems form in a completely different environment. We can look at dust clouds — which are essentially stagnant in a galaxy, science states — and some outside gravitational anomaly acting on a molecular cloud parents star formation, which in turn would lead to planetary formation resulting in a solar system like ours. This is two completely different formation scenarios, yet today's science implies galaxies and solar systems formed the same way. You cannot just weigh the galaxy as the mass of what you see and say it contains X amount of gravity. You also have to account for its speed through the universe in its infancy and the energy contained in that forward motion to determine its mass accurately.

Now none of the facts, ideas, or scenarios I have proposed are outside of the realm of today's science. I use only their facts and their numbers to come up with my theories. Yet mine seem to be so far away from theirs, it seems we're looking at different data, but that's not the case. I'm looking at the same data they are looking at, yet I come up with a completely different theory. One of us is wrong.

Chapter Seven

Quantum Physics

Don't get too nervous — I will keep this as simple as possible. Every scientist will tell you quantum physics works, but they just don't know how it works. I touched on this earlier, but let me get a little deeper into it.

Quantum physics implies that anything that can happen, did happen, somewhere in the universe or in some other universe. I believe I have debunked this notion in the time-wedge chapter, but let me try again. Obviously this scenario was put forth before the 2016 World Series, but for the purpose of this book, it still works.

One example used by science today is that somewhere in some universe the Cubs won the World Series. Their explanation for this is they believe photons, electrons, and other subatomic particles are in multiple places at the same time, and until you take a reading or make an observation, then and only then does the particle decide where it is. So science claims it is the particle that decides what is real. Even if that is the case, everyone saw the Cubs lose for the past hundred years, so those particles have already made their decisions — those photons have already been encrypted with the information that the Cubs had lost the World Series, and those photons are now streaming into space. I don't believe the particles can change their mind along the way. So no matter where you are in the universe, when that sheet of photons hits your eyes, the Cubs lose.

Also if anything can happen, why don't the Cubs win the Super Bowl or the Tour de France? Remember, anything can happen anywhere, right? The scientific explanation used to justify the Cubs winning the World Series in some other universe is what Einstein called "spooky action at a distance." To put it simply, this means if an atom is spinning clockwise over here, its sister atom is spinning counterclockwise over there, no matter how much distance is between them, even light-years. I don't know how they can test that, but I will take their word for it. The phenomenon implies that if the Cubs lose over here, they win over there, wherever there is, and because the Cubs are made of particles, and every particle has an opposite, this must be true. This is also how science comes up with the notion that if there is a good Billy in this universe, there is a bad Billy in an alternate universe somewhere. Whatever.

Now as far as every atom having an opposite sister atom in some other universe, there is no proof of other universes, just calculations and speculations, which I spoke about in previous chapters. No one can point to one, and there are no tests you can do to prove the existence of other universes. But even if there were other universes, the laws of physics would be the same. The other universes would have had a Big Bang, and galaxies would have formed the same way. And the other universe would contain the same elements as we have in our universe because if you change any of the laws of physics or change any of the proportions by even a fraction, science states you cannot make a universe. Everything has to be just right.

Another example science likes to use to explain quantum physics is the cat-in-a-box test. It goes like this: You have a box, and in that box, you have a cat and a cyanide pill. Now if that cyanide pill falls out of its container, the cat dies. If it stays in the container, the cat is alive. The man who invented this thought experiment used radioactive material hammer and a vial of poison. That gets way too complicated for the purpose of this book but the principle's the same. Now until you open the box and look in, the cat is both dead and alive. It is not determined one way or the other until you open the box and look inside. Okay, but if I open the box, and the cat is dead, when I close the box, does the cat come back to life? Of course not. The cat will always be dead. You have to get another box and another cat to conduct the experiment again. Let's take this a little further. Now if I have two boxes and two cats, does quantum physics imply when I open one box and the cat is dead that the cat in the other box is automatically alive, whether I look in or not? Or is it still both dead and alive? Or does it have any bearing on it at all? I don't know — you tell me.

Now as crazy as all this sounds, I'm not making it up. This is the science going on in today's universities. No wonder there have not been any major breakthroughs made in the past hundred years.

MRIs are also used as another example of quantum physics. They claim all the magnetic particles are everywhere at the same time and it's not until the machine takes a picture that those magnetic particles make a decision and form the image that the doctor will see. So let's say the MRI captures the magnetic image of a broken ankle. If quantum mechanics is what they say it is, does that mean in another universe the ankle is not broken? And if quantum mechanics means anything can happen until you look at it, if I take another picture, will it be of a broken toe or maybe a broken foot or something else entirely? Of course not. If you take ten pictures, you will have ten pictures of a broken ankle.

Quantum mechanics is not as crazy as science makes it out to be. What makes the calculations imply such crazy predictions is the use of time in the equation. Once you remove time from the math, quantums begins to make sense. But as long as science insists time is a real thing, they will keep coming up with ridiculous theories that even Einstein could not embrace.

As I explained in the time-wedge chapter using their cell-phone example, time is not real. Yet science uses cell-phone technology as an example of quantum physics. They say all the cell-phone signals are everywhere, and until you make a phone call, which is the same as making an observation, that cell signal can be anywhere and say anything. But if you take it down to each individual electron in a never-ending stream of electrons, once the electron has been encrypted with information, it will stay encrypted forever. As I said before, it will always say, "Hello, Frank," anywhere in the universe you receive that call.

The best proof science has today for the anything-can-happen-anywhere theory is the double-slit test. It consists of a laser shooting one photon at a time at a screen with a piece of steel in front of it. The piece of steel has two slits in it. Now you would expect the screen behind the piece of steel to show two groups of photons, one behind each slit. But that's not what happens. The photons seem to spread out and form a wave pattern on the screen. Now science states the only way this can happen is if the photon is going through both slits at the same time. But when they turn on a photon detector, the screen displays two groups of photons, one behind each slit, as you would expect. But this only happens with the detector on, and when they switch the detector off, the wave pattern returns, and it does not matter if the detector is in front of the piece of steel or behind it, which leads to another paradox that I will get to later. Science states this phenomenon is proof that the photon is in multiple places at the same time. So one photon is going through both slits at the same time, and not until you look at it with the detector will the photon decide where it is and decide which slit it will go through. So the conclusion is only by looking at the photon will you make it decide where it is and only then the photon picks one slit or the other and the wave pattern disappears. So let me get this straight — they want you to believe that the photon knows if you are looking at it or not. Really? Shouldn't the photon interacting with the atoms in the air in the room be considered a reading of some sort? If the air particles interact with the light, that means the light has established its place — or in this case, made its decision as to where it is in the room. Did I just lose you there? Sorry, but that is what science believes and teaches.

Let's try this: First let me ask, does a single photon emit light? For instance, if you put a single photon in a sealed box, does the inside of that box light up, or does it stay dark? If the inside of the box stays dark, then there is a problem — that would mean a piece of light does not give off light, and someone has some explaining to do. But if the inside of the box lights up, even a little, that would explain a lot.

I believe the photon is emitting light waves. Now the detector only detects the photon, not the waves of light emitted by the photon. So to bring it into perspective with this book, the photon is real, and the waves of light emitted from it are time. Remember, light is time, and time is not real. So when the detector is on, you only see the actual photons behind the slits, and this is reality. When the detector is turned off, the screen also picks up the light waves given off by that photon, which I propose is time and this is what forms the wave pattern on the screen

This goes back to the time-versus-reality argument. When the detector is on, the screen shows what is really happening: The photon is going through one slit or the other. When the detector is turned off, the pattern becomes dispersed because of the light waves emitted from the photon. The photons are still going through one slit or the other. The waves interact with each other, which, in this case, causes a wave pattern on the screen. So the way I look at it, the double-slit test is not a question — it's an answer. The double-slit test seems to prove my theory. I know this is a little mind-numbing, but if you think about it, it does makes much more sense than the photon knows if you are watching it or not. Just one more thing on this: If a particle has to be observed in some way or interact in some way to become part of reality, what about all those photons out there where there is no one to see it or detect it? The theory would imply that if there is no one there to make it decide where it is, then by definition it is nowhere and there should not be anything anywhere unless someone is there to detect it. Sort of like the question, does the tree in the woods make a sound when it falls if no one is there to hear it? Trust me, the tree makes noise when it falls. The laws of physics demand it.

Now another scientist took this experiment in a different direction. Instead of lasers and photons, he used a drop of silicone in a liquid medium poured over a vibrating table. Now the drop of silicone forms waves in this liquid medium, and as it moves along by the vibration, he states it looks like the waves around the drop are moving the silicone drop along. Now these waves in the medium formed by the silicone would produce a dispersed pattern on the screen behind the two slits, and he's right. I believe I just explained that, but he seems to have it backwards. The waves are not pushing the silicone; the silicone is causing the waves, and the waves are following the silicone. Again the silicone is real, and the waves are time. No matter how you do it, it seems to prove my theory. I don't understand what the big mystery of quantum physics is.

Chapter Eight

Black Holes

Everyone knows what a black hole is. It is an area in space where gravity is so strong not even light can escape its grip. But there are also different types of black holes. There are stellar black holes, which are remnants of a supernova explosion. There are also supermassive black holes, which science believes are at the center of every galaxy. And among these there are also variations. There are active black holes in the center of some galaxies, and other galaxies, such as our own Milky Way, contain dormant black holes. So what's the problem?

Well science wants to turn black holes into things like white holes or wormholes to another part of the universe or even pathways to different realms in other universes and other far-out phenomenon of the unknown. Einstein's theory implies that a black hole's mass is so great that it stretches the fabric of space-time almost to its breaking point, and it is this effect that results in these ideas. I believe I have provided enough evidence to at least argue the dent in space theory, which in itself should end such speculation, but for now, we'll go with it.

The other problem is when science tries to describe what happens to matter that falls into a black hole. Let me explain. Science tells us the math predicts that matter crosses the event horizon and continues to be pulled toward the center of the black hole. So far so good, but what happens when it reaches the center? The math tells us the matter reaches a point of infinite mass and infinite gravity, and not only that, time itself stops or wraps around on itself — it depends whom you listen to. You end up with the symbol for infinity, and with that there's no more math to do. Science calls this point in space the singularity. This is where physics itself breaks down, and science doesn't like that at all. Now this is a big sticking point to physicists because when you take the math away, that only leaves speculation, which, in turn, leads to outlandish theories, as I have pointed out. Now if you ask them about the symbol infinity, they will tell you it simply means they do not know. But if you change the word infinity to transition, that changes everything. It allows you to keep going with the math, and if you can keep going with the math, it gets rid of all the speculation and leads to a more logical hypothesis. But now you have to ask, transition to what?

To answer that question, you have to go back to how black holes form in the first place. As I mentioned before, stellar black hole formation is pretty much figured out, and I have no problem with this. It is the supermassive black holes in the center of galaxies that I want to concentrate on, and we will get to that in a minute. But first, the difference between an active black hole and a dormant black hole is, in an active black hole, it is continuously consuming the matter around it. A dormant black hole has consumed all the matter within its gravitational reach, and either the remaining matter is moving too fast for the black hole to capture it or the matter is beyond the reach of its gravitational force. Now the difference between the two is what I believe contains the key to the problem. An active black hole is spewing gamma-ray beams from its poles, some as far as 400 million light-years into space. That is four times farther than the Milky Way is across. Now this phenomenon in itself raises another paradox. If nothing can escape a black hole's grip, how are gamma rays escaping? Science has a few theories to explain this, none of which make much sense to me. They tell us that the black hole is so full that it cannot hold any more matter, so it throws the excess matter into space, the "messy eater" theory. Another theory is, the matter at the event horizon is moving so fast that some of the matter is being sucked into the black hole, but some of it is thrown back into space. Now I don't understand how, of the matter in the same place, half of it goes into the black hole and half of it gets thrown back into space, but that is another of today's theories. Now this would be probable if the matter were being thrown into space along the equator — angular momentum would kind of explain it — but that is not what is happening. The matter is being expelled at the poles. But how does the matter migrate from the equator to the poles before getting thrown in space, and how does the matter get turned into pure energy along the way? Again show me the math on that.

Now as I said in the beginning of the book, I don't intend to just pose questions, as science today does. This book is to offer logical explanations for the phenomena we see in our telescopes and detect in our instruments, so let me give it a try. Let's start with an active black hole. It is consuming matter along its equator and spewing out gamma rays from its poles. None of that statement can be disputed. But how are these two actions related to each other? I believe the black hole sucks up the matter, and this matter continues moving to the center, being pulled along by the extreme gravitational force of the black hole, but once the matter reaches the singularity, that's where the math breaks down. I believe it is transformed by the extreme heat and pressure into pure energy, gamma energy. Now even a black hole's intense gravity cannot contain an infinite amount of energy. Sooner or later, the internal force of all that pure energy trying to escape will become stronger than the gravitational force of the black hole, no matter how big the black hole is or how strong the gravitational force is, and something has to give. I believe eventually the gamma energy breaks through the event horizon at its weakest point, which would be at the poles, and this would result in the gamma-ray jets we see in active black holes. Once this breach in the event horizon happens, the gamma-ray jets will continue for as long as the black hole keeps consuming matter. But once the black hole has consumed all the matter its gravity can reach, the transition of matter to energy stops. This, in turn, lessens the outward pressure, and the black hole's gravity wins again, the event horizon closes, and the gamma-ray jets turn off, and this results in a dormant black hole. Do I make sense so far?

But I don't believe that an active black hole would become a dormant black hole just like that. There must be a phase in between that can be characterized as a semi-dormant black hole. As the matter around the black hole gets depleted, it would be consuming less and less matter and not at a steady pace, thus producing less and less energy, and the gamma-ray jets would happen at different intervals, erupting only when the black hole has reached its capacity again. This would be a semi-dormant black hole. Now this occasional eruption is what I believe a gamma-ray burst is, and even a dormant black hole will capture an occasional star or some other object that comes within its reach, and again, sooner or later, the energy would become too strong for the black hole to contain and again spew energy through its poles, but only for an instant. This too would result in a gamma-ray burst. Now science detects these gamma-ray bursts throughout the universe, but they have no explanation for them. I believe I just proposed one logical scenario that could account for this phenomenon. This theory would also explain why we see gamma-ray bursts coming from all across the universe. There are black holes in every stage of development throughout the universe. Logic and my theory would imply there is a correlation between the age of a galaxy and the amount of activity in the black hole in its center. Has anyone looked at the universe using this theory yet? I wonder what their math would come up with if they did.

This theory allows the math to continue once you get to the singularity. It eliminates the finality of their calculations and leads to a proposal of thought that can at least be debated. So by changing the word infinity to transition, you can use math to determine how much mass is being transformed into how much energy. You can use the math to determine how much energy a specific-sized black hole can hold before being breached and to answer other questions I will get to in a minute.

Science uses its own thought experiment to describe what would happen if we ourselves went to a black hole. It goes like this: Two astronauts take a spaceship to a black hole. One astronaut stays in the ship outside the reach of the black hole, and the other one takes a space walk to the black hole. Now when the spacewalking astronaut passes through the event horizon, science and the math tells us the astronaut becomes stretched out — or what science calls spaghettified — because the gravity at the astronaut's feet is stronger than the gravity at the astronaut's head. A little side note: Let's see a dent in space-time explain that. But that's not the problem. The scenario also predicts that the astronaut in the spaceship would see his fellow astronaut get to the event horizon, but he would never see his companion enter the black hole. Science claims his companion's image would be permanently emblazoned on the surface of the event horizon even though the astronaut had passed through and is now stretched to the extreme inside the black hole, their logic being time has stopped for the astronaut at the event horizon. So I guess the image must stop where its time stops. I don't really get that but whatever. I thought time stopped at the singularity. This is another paradox of the math. The astronaut is not there, but the astronaut is there. Now if their scenario is right, you should be able to look at the surface of a black hole and see everything that the black hole has consumed over its entire lifetime because all their images would be on the surface of the event horizon. Really? This sounds absurd to me; how about you? This is another product of using time in an equation.

Now let's get back to reality. First off, the astronauts, the spaceship, and anything else, for that matter, would be torn apart by the gravitational force and incinerated by the extreme heat around the black hole. The only thing that would be sucked up would be the smoke and the ashes that remain. If Jupiter's gravity can tear apart a comet, what would the gravity of a black hole do to a spaceship and the two astronauts? I do not even believe a black hole is black, at least not on the inside. Remember, to say light cannot escape the gravity of the black hole implies there is light inside in the first place. So the only part of a black hole that is black is the event horizon itself.

This theory also implies the event horizon would become thinner as it sucks up more and more matter, which produces more and more energy trying to burst out against the gravity of the black hole. The light or gamma energy would start out deep inside the black hole, so the event horizon would be very thick, and as the black hole consumed more matter and transformed that matter into energy, the force would push the inner event horizon out, making the event horizon thinner and thinner. This battle between gravity and energy continues until again there is so much energy inside that even a black hole cannot contain it and the black hole reaches its breaking point. There is all kinds of math that can be done on black holes, such as how thick is the event horizon in any given black hole? What is the breaking point of a given black hole? How much gravity does it take to stop light in the first place? And so on. But to keep doing the same math and arriving at the same infinity and throwing your hands up in the air saying you don't understand it over and over again is, by Einstein's definition, insanity. Yet this is where science is today. But by using logic, there is no need for white holes, wormholes, or the breaching of the fabric of space-time, and you would not have the need for other dimensions or universes. I believe my theory gives you just reasonable and predictable phenomenon that can be explained with simple math. Now I know this kind of ruins the TV shows and movies based on the theories science has touted as fact for so long, but to ignore logic seems irresponsible.

A recent observation of a supermassive black hole in a nearby galaxy found a direct correlation between the size of a galaxy and its black hole. This puzzled scientist even though it seemed to make sense. The question that stumped them was, how can something so small by comparison to the galaxy influence the whole galaxy? They have it backwards again. The galaxy controls the black hole, not the other way around. Which scenario makes more sense to you?

I believe supermassive black holes formed along with the galaxies. Again the conditions in this early universe were extreme, the most extreme the universe would ever see. The heat and pressures existing only this once in the universe is what would be needed for matter to go right past the star formation stage to black hole. During this time, there would be no limit, except the amount of matter available. The amount of matter in any one area of space would determine the size of the galaxy and the black hole in its center. Supermassive black holes and stellar black holes form under completely different conditions. Galaxies and supermassive black holes formed at only one period of time in the universe. Stellar black holes are forming all the time. The two have nothing to do with each other.

Chapter Nine
The Atom

We can all remember from high school science class that an atom is simply electrons circling a nucleus of neutrons and protons, except for hydrogen, the basic element, which has no neutron. The number of electrons, neutrons, and protons in the atom determines which element you have. The periodic table lays this out in a neat little package. It begins with hydrogen with one electron and one proton and ends with copernicium with 118 electrons, neutrons, and protons. There are some other elements beyond that, which have yet to be confirmed, but for my argument, it doesn't make a difference. So what's the problem? Nothing so far, but when science describes what is between the electrons and the nucleus, I have a big problem. Let me use an analogy that science uses to describe the atom. A scientist tells us if you were standing in the gymnasium of some university and you put a marble in the center of the room, which signifies the nucleus of an atom, the electrons would be circling around the walls of the gymnasium. So far so good, but in his next statement, he goes on to say between the electrons and the nucleus, there is nothing, which in turn implies an atom is made up of essentially nothing, and whereas all matter is made up of atoms, this means matter is made up of essentially nothing. This is where I have a problem.

I find it hard to believe that everything in the universe is made up of nothing. In this book, I am trying to deal with logic, and that conclusion I find very illogical. How can everything be made up of nothing? Do you see where I'm coming from? I do not believe we, or any of the other things in the universe, are made up of nothing. There must be something between the nucleus and the electrons circling around it, and something else must be going on here, but what? I believe it is energy — the raw energy from the Big Bang or, better yet, the raw energy of the atom itself. Let me explain.

One of the basic tenets of science is the Big Bang. In the beginning, energy was thrown into space — at what speed is up for debate, but for my scenario, a specific speed is not necessary. Science claims pure energy was thrown into space, and from this original energy, matter was formed. But how? Science tells us that over time, as the energy expanded, the energy cooled, subatomic particles formed, and then these subatomic particles, in turn, formed protons and electrons. Then from these electrons and protons, hydrogen atoms were formed. And again I ask how. At what point does energy become matter? Today we can turn matter into energy with no problem. We take carbon energy to drive our cars, run our factories, turn our lights on, and heat and cool our homes, and there are many other ways matter has been turned into energy since Homo sapiens first discovered the secret of fire. The very existence of the universe relies on the conversion of matter to energy in stars and supernovas. Science can take Einstein's theory from matter to energy, but no one has an explanation of the process that turns energy into matter, which is the other implication of Einstein's theory. If it is as simple as $E=Mc^2$ reversed, why hasn't some university taken some energy, whatever amount the math indicates, and cool it down to form a piece of matter. I have to say, if the science is so conclusive and the process so cut and dry, build me a proton.

Let's get back to the Big Bang. Science claims all the matter that makes up everything in the universe was formed from a bit of intense energy smaller than an atom. Science also calls this a singularity. This notion again comes from the assumption everything in the universe is moving away from us at whatever speed they wish to assign to their calculations, and by running the clock backwards, all of the matter in the universe ends up at a single point. Again by using the same math as a black hole, you end up with a singularity. But this implies that the Earth and the Milky Way are the center of the universe because science also claims everything in the universe is moving away from us. So this implies that the Big Bang happened right in our back yard and everything in the universe has expanded from here. Now does anyone really believe that the Milky Way is the center of the universe? Another problem with the math is, how do you accumulate enough energy in a single point in space to form a universe?

To try to determine how much energy we are talking about, let's do another thought experiment using Einstein's theory of energy equals mass. Remember, science tells us all the mass in the universe came from the energy released from the singularity at the instant of the Big Bang, okay? Here we go. The energy that was released in the atomic bomb that was dropped on Hiroshima came from a piece of matter the size of your fist. So how much energy would be released from a piece of matter the size of the Earth? Now how about a piece of matter the size of the sun? How much energy do you get from a piece the size of a galaxy? Whatever number you have, that's how much energy science implies you would need to form the matter in a single galaxy. Now there are hundreds of billions of galaxies in the universe. How much energy is that? But we are not done yet. This number does not even include the galaxies we cannot see, and even with that, we are still not done.

Science will tell you that after the Big Bang, both matter and antimatter were formed, and as the universe was forming, matter and antimatter began annihilating each other. But there was one piece of matter per billion more than antimatter, and that is the matter that we see throughout the universe today, just one in a billion. So to come up with the initial amount of energy in the singularity before the Big Bang, you have to take the energy it takes to form all of the matter in the universe, as we just did, and multiply it by two billion — one billion for the matter that was annihilated and one billion for the antimatter that was annihilated. How big is the number now? How many zeros do you have? What do you call that number? I believe we are well beyond a gazillion or even a googolplex by now. But whatever you call that number, that's how much energy you need to create a visible universe that we see today — and that's only in my universe. So you want me to believe that all that energy was contained in a space smaller than an atom. Again I have to say, really? And that does not even include all of the dark matter and dark energy that science claims makes up more than 95 percent of the universe. This means the number we came up with is still only about 5 percent of the energy contained in the singularity before the Big Bang. There is no calculator in the world that can come up with that number.

Now whatever it was that blew up resulting in the Big Bang, we can all agree it was a tremendously violent explosion, and energy was propelled into space much faster than the speed of light. Now with all that energy traveling at that speed, along with all the heat and pressures involved in such an explosion, not to mention all the extra heat and chaos caused by the annihilation of matter and antimatter that science claims is happening at this point of the process, I believe this is when protons and electrons were formed. How they were formed we can argue about. I believe they were formed by the raw energy concentrating into stable bits of energy, and these bits are still nothing more than energy. But once these protons and electrons formed — one with a positive charge and one with a negative charge — the electron circled the proton, and hydrogen was formed. But this still does not explain what is between the nucleus and the electron.

You have to remember this all happened as the primordial burst of energy was traveling at faster-than-light speed. Also as an atom is forming, electrons travel at the speed of light. So if an atom was a mile around, the electron would circle it 186,000 times per second. Now we all know how small an atom is. So the electron must be traveling around the nucleus — again there is no number for how many times per second. So during formation, the electron simply captured the energy that was around the nucleus, and a hydrogen atom was formed. We now have a piece of matter, but its basic components are still energy. So do we have matter or energy? I say we have both — at its basic level, matter is energy. Also, though the electron's mass is small, the speed of the electron contains energy. Also how much energy does it take to stop the electron from just flying off? There must be a force holding the atom together. Gravity holds the solar systems together. So is it basic or gravitational magnetism at work here? Energy is mass, so therefore, with all that energy between the electron and proton, the energy needed to hold it together, the energy of the electron moving, and the energy of its basic components, the atom is full of energy and, therefore, mass; we are again made of something. Now if you have one atom of iron, is it a piece of matter or a piece of energy acting like matter? How about two iron atoms? Now you have interaction between two atoms. Is this the point where energy becomes matter? Does it take two pieces of energy to make matter?

This theory would explain the inside of an atom much better than there is nothing between the electrons and the nucleus. Therefore, instead of everything being made up of nothing, my theory proposed everything is made of energy, and, remember, energy is mass. You don't need a Higgs boson to give matter mass — matter is mass, mass is energy, and as Einstein stated, they are the same thing.

Now if you believe there is energy trapped inside the atom and the atom itself is energy, as I have just proposed, it would explain the energy released in a chemical reaction. For instance, if you place a piece of sodium in water, it will fizzle and release energy. This energy must be coming from somewhere. If an atom is mostly nothing, where does the energy come from? Now if you mix sodium and chloride, you get such a violent reaction that a dangerous amount of energy is released. Again I have to ask, where does this energy come from? Now let's take it to an extreme. How about a nuclear explosion? Just a fistful of matter destroyed an entire city in Japan. Does this sound like an atom is made of nothing?

I don't want to get too deep into this, but if you look at the structure of an atom such as hydrogen, it is one electron orbiting one proton. The orbital distance of the electron is called the shell. It is at a set distance from the nucleus. Now each shell can only hold a set amount of electrons. The first shell can only hold two electrons, the second and third shells can hold eight electrons, the fourth shell can hold eighteen electrons, and so on. Each of these shells are farther away from the nucleus of the atom, which I propose means it has trapped more energy, and it takes more energy to hold more electrons together — more electrons contain more energy. Remember, atoms were formed under extreme conditions, and it requires tremendous amounts of energy — energy, I believe, that gets trapped by the electrons when the atom is formed, and this continues up through the periodic table. So it is not as simple as just electrons circling the nucleus. The more electrons, the bigger the atom is and, therefore, the more energy it contains. So what I am proposing is the more electrons you have, the more energy can be trapped by the electrons. So the more orbital spheres surrounding the nucleus means there is more energy trapped inside the atom. So the electrons circling the nucleus are not orbiting around nothing, as science states today. I believe they are orbiting around the energy that the electrons trapped in the atom and the energy in the atom itself. I cannot give you the math that shows this, but science can't give you the math that explains energy cooling into matter. But I hope you get the gist of what I am trying to explain. My point is, I do not believe that there is nothing inside of the atom. And today's theories are just that — theories. They have no more proof for their theory than I do. But I have much more logic in my theory than they do. Who is right? Well that remains to be seen.

Chapter Ten

Particle Physics

For over fifty years, we've had particle accelerators smashing protons to-gether trying to reveal the basic building blocks of matter, the theory being, analyze the pieces that result from breaking apart a proton. In other words, what is a proton made of? But when they began smashing protons together, it seemed that there were hundreds of different subatomic particles, which they called hadrons. This collection of subatomic particles resulted in what was called the particle zoo. But there were so many different subatomic particles, scientists knew it could not be right; there could not be that many basic pieces, so they had to do something. They decided to narrow the list down to sixteen subatomic particles with names like quarks, leptons, and bosons. Then they basically just dismissed all the others as if they didn't exist. What gives them the right to pick and choose? It is either a subatomic particle or not, and as time went on, they added subsets of these basic particles, such as a muon, tau, gluons, and pions, with even more subsets, such as up, down, and even a charm quark. No, I am not making this up. They are very serious, and they really be-lieve this stuff.

Let's go back to the beginning to the Fermi Tevatron in Chicago, one of the first colliders. They never seemed to get the results they were looking for. In fact, on their own website, they stated they got no significant results for the first thirty-five years, and not until talk of cutting back or even closing down the program did they begin to get results. They finally began to find the subatomic particles they were looking for. Suddenly there were subatomic particles everywhere, so many that they didn't know what to do with them. But even with all this, they couldn't find the ultimate particle: the Higgs boson, otherwise known as the God particle. So their answer was to build a bigger machine, thinking if they increased the intensity of the collisions and made smaller and smaller pieces, they would eventually find the elusive boson. This resulted in the Large Hadron collider in Cern, Switzerland, the largest collider in the world, and on July 2012 they put out a press release stating they "think" they have found the Higgs boson. But only with time, more power, and many more years of research will they know for sure. It seems to me they have a predetermined outcome and will do or say anything to achieve it.

I know I'm getting a little deep here, but it should get clearer as we go on, so hang in there. When they smash those protons together, they believe they are breaking the proton down into its fundamental building blocks. I believe they are simply transforming matter into energy. When a proton is smashed, the so-called subatomic particles that are the result only last a tiny fraction of a second. This hardly seems like a particle or a piece of matter to me. I believe what you have when you smash a proton is simply bits of energy decaying. Naturally some of these pieces of energy would be bigger than others; some will go up, some will go down, some will go left, and some will go right, and most of the energy will simply go straight ahead. This is exactly what you see when you look at the pictures put out by their website. To me this is exactly what you would expect, but to scientists, each one of these streaks of energy is a different piece of matter, and what direction it flies in is the name of the particle, such as an up quark or a down quark. My question is, how can something that only lasts a fraction of a second be considered matter? Furthermore, science seems to believe that if you smash the protons together even harder, you will get different subatomic particles. I believe you just get smaller pieces of energy. And the more energy you use to smash them together should result in more energy that is decaying overall. This does not mean you have a heavier subatomic particle; you just have more energy. Have I lost you yet?

While science was looking for a certain piece that they predicted would be in their collision, a subatomic particle they called the W boson, they searched one billion collisions and came up with five examples of it. Let me use what I call the broken-glass analogy. If I predict that when I break a glass there should be a perfectly square piece or a perfectly triangular piece or whatever, and it takes me breaking a billion glasses to come up with five perfect square pieces or five perfectly triangular pieces, does that prove my hypothesis? Let's do a little math. Say I break a billion glasses, and let's be conservative and say each glass breaks into one hundred pieces. That's one hundred billion pieces of glass. If I find five pieces that fit my prediction, does that mean I was right? In any other scientific discipline, that ratio would be considered insignificant. If something is there, it should be there most of the time, not just five per one hundred billion. Another example would be if I predicted potatoes look like Abraham Lincoln, and I have to look at one hundred billion potatoes to find five that resemble Abraham Lincoln. Does that prove my prediction? I would be embarrassed by results like that. But that is exactly what particle physicists seem to be doing. Again I am not making this up. Five W bosons per billion collisions is their number, not mine.

Now back to Einstein. Humans have been transforming matter into energy for as long as man has been on this earth; there is no new science here. Cavemen burned wood to produce heat, kerosene was burned to produce light, and gas is burned to drive our cars and heat our homes. Turning matter to energy is simple. Now if you really want to impress me and prove your theories are correct, take all those bosons, leptons, quarks, and muons, etc., and build me a proton. Run that collider in reverse. Take some raw energy and make me a proton — take energy and convert it into matter. That would be something.

Science tells us that the Big Bang was pure energy that exploded to give us our universe, and as it cooled, energy was turned into matter, and that matter eventually became hydrogen, which became the basic building block of every-thing we have in our universe today. Well if the science is so conclusive and you have your little particle zoo that contains all the elements in the recipe, let's see someone do it in the lab. Smashing matter into energy is child's play. I am waiting for someone to take some energy and build me a hydrogen atom. Until then I simply believe they're wasting their time naming all those bits of energy.

Let's go back to the drinking-glass analogy for a moment. If you smash a glass against a brick wall, you end up with a bunch of pieces of glass, some big, some small. Now you can take those pieces of glass and name them if you'd like, but they are just pieces of glass, no matter what you call them. Now if you smash a hundred glasses against the wall, you will probably get some pieces that resemble each other. If you take all the pieces that look like each other and put them in groups and give these groups names, that does not mean you have some new element — you just have pieces of glass placed in groups with different names. Now when you smash a proton into pieces, you don't have new particles — you just have pieces of protons. Now to put the glass completely back together, you need all the pieces, not just the ones you named. You need the tiny pieces of glass and even the glass dust, nothing can be dismissed, yet when smashing particles together, science disregards some of the pieces of energy completely. To them only certain subatomic particles count.

This is pretty much what science is doing today. The only difference is, with the pieces of glass, I can hold them, look at them, and examine them, whereas the proton pieces only last a fraction of a second. It's not like they can take that piece of proton and examine it and do research on it. Furthermore, they then look at the universe and try to find these subatomic particles even though in the lab they only last a millisecond, and even if there are subatomic particles, they don't last long enough to do any research on. How can something that only lasts a fraction of a second be responsible for all the mass and all the matter in the universe? Science also talks about subatomic particles popping in and out of existence. Now when it pops out of existence, would that not be losing information? Remember, no information can be lost, according to the rules of science. This is another paradox that seems to be ignored. I hope you were able to follow this. If not, all you really have to know is science believes everything in the universe is made up of subatomic particles that last only fractions of a second, and the Higgs boson, which gives all matter its mass, is so elusive they are still not sure if it's real. Does this sound like real science to you?

Chapter Eleven

Light Speed

Nothing can travel faster than light. This has been a basic law of physics since Einstein came up with his theory of relativity over a hundred years ago. But again this leads to another paradox, which I will get to in a minute.

But first, the basic premise is that as you approach light speed, you gain so much mass it would take an infinite amount of energy to propel you that extra bit to reach light speed because you have become infinitely massive. There is that word again: infinite. There is no such thing as infinite. It is simply a product of math, playing with numbers and using time in their equations. If the answer to your equation is infinity, you have the wrong answer or the wrong equation. Yet in the early 1960s, particle accelerators seemed to prove this; they could only get the particle to go so fast. This may be so, but the problem is not that the particle is gaining mass, the problem is the propulsion system. Seems you would need a force faster than light to propel something faster than light. Remember the BB analogy we went through before. The BB does not get any bigger — it just goes faster — and the energy it contains from the speed is the extra mass. It's not that the BB weighs more or grows bigger; it's always the same size. In particle accelerators, they send protons around the circle at 99.99 percent the speed of light. Does the proton get excessively massive? Does the proton grow any bigger? Science says yes; I say no — it stays the same size, and it just contains more energy. So why would an atom or even a spaceship grow to some extreme size that it could not reach light speed? The limit is engineering, not excess mass.

Now for the paradox. After the Big Bang, science tells us stars began forming 200,000 years later, and then one million years after the Big Bang, galaxies formed. No problems so far. Now let's take a look at the Hubble deep sky photograph. It is a picture that shows infant galaxies 13.7 light-years away in one direction. Now we can assume there are infant galaxies 13.7 light-years away in the opposite direction. That puts infant galaxies 27.4 billion light-years apart from each other. Yet science will tell you they are only one million years old. How did they travel 13.7 billion light-years in just one million years without faster-than-light speed travel? Again these are science's numbers, not mine. These are facts accepted by everyone. This also assumes the Big Bang happened right here in the Milky Way. If the Big Bang happened somewhere else, the numbers and the paradox get even worse for light speed being a limit. The data shows this has happened, yet science will tell you it is not possible, yet there it is on film. So to explain the contradiction, they tell us the space between the galaxies was expanding faster than light, so the galaxies are traveling with space faster than light, not traveling through space faster than light. Right, that explains it. This leads to another difference between my theories and today's science. They believe the Big Bang created space; I believe the Big Bang happened in space. This is why I believe new breakthroughs have not happened in science in the past hundred years. You have to question conventional beliefs to move forward in science. You must think outside the box. If we hadn't, the sun would still be going around the Earth. Yet today's scientists want to stay in their comfort zone, so to speak, just doing the same experiments and using the same math and expecting different results. Again Einstein has a definition of that.

One scientist at MIT came up with another theory to explain this; he calls it cosmic inflation. I won't get into it too deep, but just let me say, even other scientists cannot make sense of it, never mind normal people. It claims for just a fraction of a second after the Big Bang, the universe expanded greater than the speed of light and grew exponentially, so a few billionths of a seconds worth of faster-than-light speed expansion explains everything. Does that make sense to you? But even if he is right, which I doubt, it means faster-than-light speed is possible. Do you see what I mean? Science wants it both ways. Cosmic inflation implies faster-than-light speed travel, but nothing can travel faster than light. So to justify the paradox, they tell us that there was nothing in the universe at that time — no matter had formed yet. So what was traveling faster than light was nothing. Well that explains everything, right? I am not making this up, and they are very serious.

So even when science proposes a new idea, it seems to come right out of *Star Trek* —absurd theories to explain simple things, or theories that cannot be tested, so no results are possible. It seems to me science today cannot make any real headway, so let me give it a try.

Until Chuck Yeager broke the sound barrier, many scientists thought it could not be done, or if it was broken, it would lead to catastrophic results. They were wrong on both counts. The sound barrier was broken, and nothing except a sonic boom happened — hardly catastrophic. So now I ask, why would the light barrier be any different? Sound and light are both waves, so the same rules and laws should apply to both.

Think of what happens when you break the sound barrier. The faster you go, the more compressed the sound waves get until they are flattened into a thin disk or barrier. Then when the jet reaches a certain speed, the jet flies through the disc and breaks the sound barrier. Now once the jet is beyond this disc of sound, the jet seems to make no sound at all. But of course, it still does. The sound of the engines is now behind the jet. So when you look at the jet, the sound is no longer coming from where you are looking; it is coming from behind the jet somewhere, depending on how fast the jet is going. Why would light waves act any differently?

Let's go back to Einstein on his train traveling at the speed of light. If you were observing him from the side, you would see a flat disk of light going by. Today's scientific belief states Einstein himself would be flattened to a disc. But this belief is also wrong. Again you are not seeing Einstein, you are seeing the light bouncing off him. Einstein is not flattened into a disc, the light bouncing off him is. This is another misconception of Einstein's theory. Einstein is real; the light bouncing off him is time.

Now let's apply my theory to the scenario. As the spaceship approaches light speed, the spaceship does not compress as science implies; it's the light waves bouncing off the ship that do, just like the sound waves of the jet. Now as the spaceship accelerates even faster and passes light speed, it pierces the disk of light and flies through it, just like the jet going through the sound-wave disc.

As an observer on the ground, you would see the disc of light but not the spaceship; it is now in front of the compressed light disc, and for all intents and purposes, the spaceship is now invisible. Just like you don't hear the jet, you don't see the spaceship. Am I making any sense to you? Light speed is not a limit of speed, it's a limit of sight. You cannot see anything traveling faster than light. Because you cannot see something does not mean it's not there. This would also explain the size-of-the-universe paradox, and it makes faster-than-light speed travel not only possible but necessary to explain the universe we see today. If the universe can travel faster than light, then other things should be able to also. The limitation is not the laws of physics, it is the energy source. What do you think? We will get deeper into this later.

Science also uses things like wormholes and bends in space-time to beat the speed of light. The next chapter will deal with this, so I won't get into it right here.

Now it's my turn to get a little crazy. Try this. Let's say Einstein on the train goes faster than light and looks back at the light disc of himself traveling on the train. Is he missing from the image because he is now in front of the light disc, or does he see himself as he was a few seconds ago? Think about that for a minute. But even in this example, it is still not time travel. Even if he was going twice the speed of light, once he began to slow down, the light or time would catch up to him, and everything would be back to normal except for the time he missed while he was on his journey. As I keep saying throughout this book, the time-wedge theory eliminates many of the paradoxes in today's scientific theories and gives science new ways to look at things. Without these paradoxes, science can finally move onto much more logical theories and areas of research.

Chapter Twelve

Time Travel

I believe I already covered this in the previous chapters, but I think it deserves its own chapter just to get into it a little deeper. Ever since Einstein proposed the possibility of time travel, science has been trying to come up with ways to make it possible. Dozens of theories and scenarios have been proposed to make time travel more than a theory. Most of them require near- light-speed travel or gravitational forces acting upon space-time. For instance, the closer you get to the speed of light, the slower your biological clock runs — you literally age slower. Gravity also affects time. The closer you are to a gravitational field, the slower a clock moves. This part of the theory is true; this phenomenon really happens. Today's atomic clocks are so accurate that these ideas can be tested and proven. There are experiments that have been done, and the results leave no doubt, but none of those phenomena describe actual time travel. All of them are just time manipulation — special effects, if you will, in the movie of reality. A supermassive black hole is the most extreme example of gravity. Its gravity stops time, but yet throughout the galaxy around it, time goes on. Stars form and die, solar systems form, and on a tiny planet in one solar system twenty-six million light-years away from the black hole, life goes on.

True time travel is going into the future or past and then returning, like in the movie *Back to the Future*, where Marty went back in time to 1955 and then returned to 1985. The movie implied Marty went back to an Earth some-where thirty years earlier. I ask again, where is that Earth in space he went back to? The Earth has been moving through the Milky Way for thirty years. It is not where it was thirty years ago. How many Earths are there? In *Back to the Future II*, both Doc and Marty went thirty years into the future. What Earth did they go to in that case? There is nothing where the Earth is going to be thirty years from now, at least not another Earth. So exactly where do they go? Now when the movie ends, everyone is back where they started, but the whole situation has changed. The Earth is where it's supposed to be, but somehow all the people and all the events on that Earth that he left are now completely different. Now I have to ask, where is the Earth that he left in the first place? What happened to all the people who were on that Earth? Exactly when did the Earth of 1985 switch? Was it a gradual transition, or did it happen in the blink of an eye?

All of these theories proposed by science today to go into the future don't really describe what happens in the movie. As I explained in earlier chapters, science only manipulates time, such as when an astronaut leaves Earth and circles a black hole. The tremendous gravity slows time for the astronaut so he does not age as fast as everyone else back on Earth, and when he returns to Earth, everyone he left is either old or dead or whatever, depending on how long he circled the black hole, as well as other factors. This is not traveling into the future; this is just missing what happened while you were gone. You simply manipulated time on a personal scale, unlike when Doc and Marty left 1985, went to 2015, and then came back — no one even noticed they were gone. Their time travel affected everyone back on Earth. Situations changed for people who had nothing to do with time travel. Poor people got rich; people had different jobs. The whole world changed because two people traveled time. The scenarios science puts forth do not achieve this. Just because gravity or excessive speed or both can ma-nipulate time for the astronaut, only the astronaut is affected by this phe-nomenon. Back on Earth, nothing is changing — everyone and everything is going on as it should. Just because he was manipulating time for himself, the events happening in the universe for everyone else went on as usual. Time travel in these terms is on a personal level, just like Einstein on his train moving at the speed of light. He cannot affect anything in history or in the future in any way.

Another twist on time travel science likes to put forth is the idea of warping space-time. We have discussed my feelings of space-time, but to make a point, let's say there is such a thing. Science claims that with enough gravity, we could bend space-time enough to fold two edges of the universe like a newspaper, so you could travel from edge to edge without going across the whole newspaper. Now the largest source of gravity we know of is a supermassive black hole, which we know resides in every spiral galaxy. Yet every galaxy we observe is flat. So from our observations, we can conclude the largest source of gravity we know of cannot even bend a galaxy, not even a little bit, never mind a universe. This fact also lends credence to my notion that there is no space-time to warp in the first place.

Another theory of time travel science likes to use is a wormhole. This is another product of mathematics Einstein is responsible for. Science tells us we could enter a wormhole here and instantly end up there, wherever there is. But again we would need an infinite amount of energy to achieve this. They do not even describe how energy can open a wormhole in the first place, let alone keep it open, or how they even work. But that does not seem to stop them. It is still an accepted branch of scientific study. It also seems to me that if a bend in space-time or a wormhole were out there, we would see it. Surely if space-time were bent enough to skip from one side to the other, we would see it. We have mapped the universe in intricate detail, and no sign of any such contortion of space has ever been observed.

As far as going back in time and changing what already happened, I don't even know where to begin. Take the time-machine scenario. Let's say you wanted to go back and stop the Titanic from sinking. What Atlantic Ocean do you go to? On Earth today, the Titanic is on the bottom of the sea. Let's say you did somehow go back in time and did stop the Titanic from sinking. Back here on Earth, would the Titanic suddenly rise from the sea in front of us? Would all the people who died somehow come back to life? Would Robert Ballard and Alvin, the deep-sea probe, suddenly be a waste of time? Exactly how does this work? To think we can go to the future and experience something that has not happened yet, or to think we can go back in time and change something that already happened, is just a good movie plot, nothing more.

Time travel is a theory based on a false premise that time is a real thing. Science is going to have to come to grips with the fact that it is nothing more than a movie of what already happened. I know I am beating a dead horse here, but it is just very frustrating to keep hearing that true time travel, like Doc and Marty, is a possibility. History is history, and the future is the future, and nothing we do here in the present can affect either. Things can only happen once, and when it happens, it is done; you don't get a redo. No second chances. The one thing all today's scientists have in common is *Star Trek* and *Star Wars*. They grew up with it, and in school they were taught that time travel is a possibility. They want it so badly they will do or say just about anything to justify their belief, even denounce reality. I realize that many things that were once considered science fiction are now science fact. For instance, Jules Verne's book *From the Earth to the Moon* was surprisingly accurate in its details. His location was the west coast of Florida; NASA launches from the east coast. He had the launch speed and trajectory just about right. He described retro-rockets and returning by parachute and many more similarities to today's space programs, all of this well before space travel was even considered a reality. The difference between *Star Trek* and Jules Verne is Jules Verne took science and wrote fantasy. Today's science is trying to take fantasy and turn it into science. They have it backwards, and this is why real progress in theoretical science will never happen.

Let me put the final nail in the coffin of time travel. I call this the ultimate paradox enigma, the grandfather paradox on steroids. I take my time machine, go back in time, and kill Einstein before he comes up with his theory that leads to time travel. Think about that one for a while.

Chapter Thirteen

String Theory

Now this one really drives me crazy. If you want to talk about science gone mad, this is it. Here we go.

The basic premise of string theory is that all particles are made up of strings, and these strings vibrate just like they do on a guitar. Also as you can get different notes on a guitar by making the strings vibrate at different intervals, you get different elements by having these strings vibrate at different intervals. A little faster is one element, a little slower is another element, and so on. So basically all of the elements on the periodic table are made up of the same thing, but the strings that they are made of are vibrating at different speeds. No, I am not making this up.

I don't know what this theory does to the periodic table. So now the amount of protons and neutrons in the nucleus and the amount of electrons circling it mean nothing? The element is now determined by the vibrations of the strings that are contained in the element, not the atomic weight or the construct of the atom itself. I don't get it. First off, what makes the strings vibrate consistently to keep the element stable? What makes them vibrate in the first place? Is there a new source of energy no one knows about? What determines the intensity of the vibrations? What causes the difference in intensity to form different elements? If the string slows down, does this turn one element into a different element right before your eyes? If they are right, you should be able to manipulate the basic energy source, whatever it is, to produce any element you want — a form of the philosopher's stone. It just seems a bit ridiculous to me.

String theory also requires ten dimensions plus time. That's right, to make the math work, you need the three dimensions everyone knows about, plus seven other dimensions, plus time, for a total of eleven dimensions. I thought the theory of everything was supposed to be simple and elegant. But it gets even worse. To explain why we cannot detect these other seven dimensions, science claims they are rolled up like little tubes so small that we cannot detect them. They use an ant crawling down a paper tube as an example of these other dimensions, stating the ant in the tube cannot discern any of the other dimensions outside the tube. So we are the ant in the tube, and we cannot see outside to detect these other dimensions. Well that explains everything. Even if that is the case, don't each of these tubes also have three dimensions of their own, no matter how small the tubes may be? Each of these seven tubes would have a top, a bottom, and sides. So now we have the three dimensions we know of and the three dimensions in each of the seven tubes of the other dimensions, which brings the total of dimensions to twenty-four plus time. That brings my count to twenty-five dimensions. It's beginning to get a little out of hand, isn't it? And one more thing about these strings is the size of them. Science tells us if an atom were the size of the universe, a string would be the size of one tree in your back yard. Again I'm not making this stuff up. And not only that, all of these other dimensions are supposedly right under our noses. It's not bad enough that string theory requires ten physical dimensions plus time — it gets even worse. The string is only one part of the theory. The theory is also known as M theory, which stands for a few things, such as membrane or the mother of all theories, among other names. They cannot even agree on what M means. But it goes something like this.

Imagine the universe as a membrane similar to a sheet of paper, and there are other universes, which are also necessary for the theory to work, and these other universes are also like sheets of paper. Now these membranes are separated by a tiny space between them, and these membranes undulate like waves on the ocean, and occasionally these waves in the membranes touch each other, and when these two so-called universes touch, they cause an explosion, and this explosion was the Big Bang. This is their explanation of what was before the Big Bang. But the whole premise is based on the universe being a flat plane. That is like starting out a theory like this: Imagine the sky is red. Well everyone knows the sky is not red. So the theory dies right there, and the universe is not flat like a piece of paper. Even if it is, the membrane is at least 24.7 billion light-years thick because that is how far our visible universe is from one side to the other or top to bottom. We can see that far in each direction. It is not debatable, it's a fact. That is one heck of a piece a paper.

This is a perfect example of science and math gone crazy. If you have to add seven dimensions that are all rolled up so small you cannot detect them, then assume the universe is flat and there are other flat universes close enough to touch, but not close enough to see, and then you have to add time to make your theory work, it seems to me you are trying just a little bit too hard. Also if the universe is a membrane, what does that do to Einstein's dent in space theory? Wouldn't all the dents of gravity in space-time make the membranes touch each other just about everywhere and cause multiple other Big Bangs? If string theory is right, there would be Big Bangs all over the place. They would be happening everywhere, not just one every fourteen billion years or so. But up till now, there has only been one we know of.

The whole notion of string theory, membranes, and other dimensions is another example of a theory that cannot be tested. Scientists themselves tell us a string is so small that they will never be able to be detected. So how can this theory even be considered viable? It's just conjecture, requiring things that do not even exist and conditions that are purely hypothetical. Back to the rules: If a hypothesis cannot be put to the test or observed, it is not science.

Just one more thing about this theory. One of the scientists working on string theory scrapped fourteen years of research in another area that he was working on prior to string theory to work on this idea. His previous research was completely fruitless. I believe in another fourteen years, he will probably scrap this theory too. Also who paid him for that fourteen years of work that ended up being essentially useless, and who will be paying him for the next fourteen years of searching for something that he himself says cannot be detected? Nice work if you can get it.

Chapter Fourteen

Is Space Curved?

Another question puzzling science today is whether space is curved or not. I am not even sure what they mean by that, but let's go with it any way. If they are asking if space is round, logic would say it is. After all, atoms are round, planets are round, stars are round, and galaxies are round, so it would make sense that the universe is also round. The problem is, as of yet, no proof has been found either way, despite science looking for it for years. They've tried telescopes, satellites, and lasers, and with years of observation and countless calculations, they've come up with nothing.

Part of the problem is they are steadfast in their belief that the universe is about twenty-eight billion light-years across from end to end or up and down. That's fourteen billion light-years one way and fourteen billion light-years the other way. The problem is not that they believe that there is nothing beyond what we can see. It's what they claim is beyond that point. There is a Russian scientist who works for NASA who has found something that may shed a little light on this question. His observations have found that at the edge of the observable universe, all the galaxies seem to be moving slightly in one direction, let's say to the left. His explanation of this is that those galaxies are falling into another universe. He proposes the galaxies are being sucked up by a second universe's gravity. Other scientists have proposed that this is why gravity is so weak in our universe. They imply gravity is super strong in this other universe, so it is super weak in our universe. Again I have to say, really? I'm sure the galaxies are moving to the left, but I have a problem with the other-universe theory.

Try this: What if that slight left turn the galaxies are taking is an arc in the orbit of those galaxies, and they, as well as us, are orbiting around the center of the universe? After all, an atom is electrons circling a nucleus, a solar system is planets circling around a star, and a galaxy is stars circling around the galactic center. So doesn't it make sense that the universe is galaxies circling around the universal center. If you go with today's theory of the Big Bang, which is when you wind back time, all of the galaxies merge into a single mass. But by using that theory, the center of the mass is the Milky Way. After all, the galaxies are supposedly moving away from us. So if you run the clock backwards, they are all moving toward us and meet at the Milky Way. Now I don't believe for a second that we are at the center of the universe, but if you use today's theories, that's what it implies. I believe I have debunked this notion in the chapter on dark energy. I also have proposed that all the galaxies everywhere in the universe are moving at more or less the same speed, not moving faster the farther away you get. So now we have to ask, where is the center of the universe, and where is the edge of the universe? Theoretically they should be somewhere in space where you could get far enough away, turn around, and look back, and you should see the Big Bang happening — someplace in space where the light from the Big Bang has not yet reached. Science believes space was a product of the Big Bang. I believe space was already there, but it was empty. The Big Bang simply filled space with matter, not necessarily completely, but how full we may never know.

But let's get back to the size of the universe. If you take that motion of those galaxies 13.8 light-years away as the arc of an orbit and extrapolate that arc into a circular orbit around the galactic core, how big is the universe now? If it takes fourteen billion light-years for a galaxy to move enough for us on Earth to detect the arc of the orbit, how big is the orbit?

I believe the Big Bang pushed matter hundreds of times faster than light speed, and that is why we cannot see the center of the universe. Just like we are two-thirds away from the center of the Milky Way galaxy, I believe the Milky Way is at least two-thirds away from the center of the universe, and all the galaxies we see are orbiting around the universal center. That would account for the galaxies we see at the edge of the visible universe taking a left, so to speak. That's where I believe science should be looking for a curved universe. Those galaxies are not falling into another universe; they are orbiting the universal center and not on a flat plane, like a spiral galaxy, but as a sphere, like an irregular galaxy or a star cluster. If there is a supermassive black hole in the center of each galaxy, which has been proven, imagine the black hole in the center of the universe. Do we even have a word for that? What's bigger than supermassive? Just like when a star explodes, there is a mass left in the center. I believe when the Big Bang occurred, some kind of mass must have been left at the point in space the Big Bang occurred, and if everything is slowing down, as I proposed in this book, it implies that eventually everything will someday be sucked back into the center of the universe, and eventually another Big Bang will happen, and it starts all over again. Science calls this the bouncing-ball theory. And if that is the case, and the universe is cyclical, which cycle are we in now? Who is to say we are in the first Big Bang? Maybe we are in the second or the fifth Big Bang, and how long is each cycle?

Now I don't believe I am proposing some wild and crazy new theory, at least not as crazy as some of the other theories out there. I'm just using the facts of science known today and looking at them logically. I'm just trying to make sense of the facts as we know them and apply the basic laws of physics without introducing far-out factors into the equation. The answers should be neat, simple, and logical, not complex and hypothetical.

Chapter Fifteen
Teleportation

Another theory in today's science that seems completely out of this world is teleportation. This theory comes right out of *Star Trek* literally, but believe it or not, I think this one does have some merit. Neither the laws of physics nor the laws of reality make it out of the realm of possibility. The idea is that you can break something down to basic atoms and then send those atoms somewhere and reassemble them exactly as they were.

I am sure someday in the future we will be able to break down things into their basic atomic structure and even send those atoms somewhere via laser beam or plasma beam or by some other mechanism. But to reassemble them exactly as they were, that is going to be a stretch, but I will even entertain that part of the process — not in the near future, but maybe someday. Who knows what technology will be available in the future? A hundred years ago, no one expected satellites circling Saturn or a spacecraft leaving our solar system. So teleportation may work someday. But not on people. I believe it could work on a rock or some other inanimate object, but how do you atomize thoughts, knowledge, memory, or even involuntary functions of the human body, like heart beating or breathing, basically the software of a living thing?

On *Star Trek*, Captain Kirk and whoever would get beamed down to some planet, and when they got there, they were reassembled, and everyone was just as they were when they had left the Enterprise. But how do they know who they are or remember why they were beamed down there in the first place? Does each atom of brain matter retain the information it contained through the teleportation process, then somehow when it all goes back together, all the information just keeps flowing, and everyone's heart just keeps on beating, lungs still breath, and so on? You see where I'm going here?

If I have a car running in the parking lot, and I teleport it across the parking lot to the other side, is it still running when it gets reassembled? Or do I have to go turn the key and start it up again? If I teleport a squirrel to the other side of my back yard, will it still be alive when it re-metabolizes? Or is it just a dead squirrel lying in my back yard? Can you atomize life itself?

Today's science is working on sending a photon from one place to another in a lab, and it seems to be working, sort of. What they are sending is a replica of the photon, not the photon itself. I won't go too crazy explaining it right now — again that's for another book. But it's quite complex, and it requires three photons to send one photon. Science is teleporting the information about a photon to another place, not the actual photon itself. They seem to be cheating again by using quantum physics instead of reality. Sending quantum information of a photon is not sending a photon. If the Enterprise used this method, Scotty would be sending a replica of Captain Kirk to the planet, not Captain Kirk himself.

That's about as deep as I can to get into it right now without going into scientific jargon and abstract theories. Take my word for it, they are working on it. So my conclusion is teleportation someday may be possible but not anytime soon, and as far as teleporting living things, no way. You may be able to teleport the hardware but not the software.

I believe the research in this area can and will someday bear fruit. Imagine that, sending atoms instead of photons along a fiber-optic line to a different place on Earth or on a laser beam to the moon. I don't believe this will happen anytime soon, but you have to start somewhere. I know this has nothing to do with time, but I thought as long as I'm writing about scientific theories, why not throw this in.

Chapter Sixteen

Antigravity

One of the Holy Grails in science is antigravity. Now you're probably thinking I am going to tear this apart too, but strangely enough I believe it is possible and even probable. Not only that, I believe I know how to achieve it. Let me explain. What I believe is possible is not antigravity, so to speak, because gravity is a basic force, not an effect. I don't believe it can be canceled out, but I do believe it can be overcome — there is a difference

My theory should be called excess gravity, not antigravity, and it goes like this: The Earth's gravity is what is known as 1 G. So if you could somehow get, let's say, 1.1 G of force above an object, theoretically you could make it rise because there would be more gravity above the object than below it — that object would be pulled to the strongest gravitational field. Now this seems obvious, but how do you do it?

First, we have to go back to Einstein: energy is mass, mass is gravity. So it seems to me to overcome gravity, you simply need enough energy, which, in turn, is mass, which, in turn, is gravity, to overcome the force of the gravity of the Earth. So where'd you get that energy. Try this: You take a circular tube with electromagnets encasing the tube, similar to a particle accelerator. Then take mercury, a liquid metal that reacts to an electromagnetic field, and put it in the tube. Now get that mercury circling inside the tube using the electromagnets to propel the mercury. This technology is used every day in amusement park rides, magnetically levitated trains, and even electromagnetic guns. You could get the mercury circling in that tube at the speed of electromagnetism, and this would produce a tremendous amount of energy contained in a neat little package. Now the only trick would be to direct that energy where you wanted it — let's say above the ring, much like the carrot on a stick in front of a horse. If you could create enough energy and project it above the object, the object would then be drawn to the greater gravitational field and, in theory, rise off the ground. How am I doing so far?

Now if you can figure out a way to direct that contained energy above you, then you should be able to direct the gravitational field you created to the left or right, forward or behind you. This form of propulsion, for lack of a better word, would also be comparatively quiet as compared to other propulsion techniques, such as rockets and the like. Also the farther away you got from the Earth, the less gravity you would have to overcome, and you would require less and less energy to go faster and faster, seemingly without limits. You would also be able to hover in place simply by controlling the direction and intensity of energy in the gravitational field, and just like the horse never catches up to the carrot, the craft never catches up to the gravitational force. So you could move quietly as slow or as fast as you wanted in any direction and accelerate instantly to great speeds and make instant changes in direction theoretically forever. Also one more thing: Shape does not matter — aerodynamics does not apply here. I think I just described a UFO.

Now what are some of the traits of so-called UFOs that are reported by people who have claimed to see them? One common maneuver is an abrupt 90-degree turn. Another is abrupt starts and stops. But making these maneuvers brings up the problem of inertia, where the ship turns quickly or stops instantaneously — the people inside smash against the walls of the craft. To overcome this, science claims you would need some sort of inertia-canceling device. But with my theory, you are in, and therefore part of, the gravitational field, so you, your insides, your coffee cup in the cup holder, and anything else in the craft are all engulfed in the gravitational field that is producing the abrupt turns and stops. That force acts upon everything inside the craft equally so you are not affected by the inertia — you are part of the motion. The gravity you have created pulls you and everything inside along with the ship. No inertia-canceling device is needed. Are you still with me?

This theory also eliminates the weight factor. The Earth pulls at 1 G, whether it's pulling on a pound or a ton. So anything over 1 G above an object of any size will lift whatever it is from the Earth, no matter how much it weighs. And why stop at 1.1 Gs? Why not 2 Gs or 10 Gs or 20 Gs or even more? What are the limitations? This technology would allow for interstellar travel because speed and force are no longer dependent on fuel for propulsion, and speed seems limitless to me. Once in space with no outside gravitational forces acting against you, less and less energy is needed to overcome gravity. So there would be no limit to the speeds that can be reached, and because there are no outside forces to overcome, and since you are not limited by a propulsion system, even faster-than-light speed should be possible. I told you we would get to that. How did I do?

Furthermore the technology to do this is already available. The hardware, the software, and the materials already exist. If you are looking for warp speed and interstellar travel, I believe I just laid out a feasible path to get there. Nothing exotic required, just simple logic and technology that exists today. Sure the details have to be worked out, but the basic concept is within the realm of today's science and technology. We don't have to wait around for some new discovery or alien technology or try to figure out a way to literally warp space or curl space-time around back on itself.

Now this might sound a little crazy to you, but it's no crazier than dark energy or string theory is. At least the alternate theories I put forth in this book seem plausible and make sense, whereas even the proponents of string theory state themselves that it can never be proven. Yet they will work on it for the next hundred years or so, and to what end? I don't know.

Well that's my excursion into science and the realm of fantasy. How did I do? I hope this book will open new doors in science and lead to research in things that science today has not even thought of. I know that statement kind of contradicts the whole theme of this book, but it is what it is. I guess that's my paradox — live long and prosper.